LUZ AMANDA AGUIRRE FLOREZ
LUDWING ROALD FLORES QUISPE

DAIRY PRODUCTION

LUZ AMANDA AGUIRRE FLOREZ
LUDWING ROALD FLORES QUISPE

DAIRY PRODUCTION

ELABORATION OF YOGHURT WITH ISOLATED FIBRE FROM QUINOA, KIWICHA AND CANIHUA BRAN

ScienciaScripts

Imprint

Cover image: www.ingimage.com

This book is a translation from the original published under ISBN 978-620-3-87275-0.

Publisher:
Sciencia Scripts
is a trademark of
Dodo Books Indian Ocean Ltd., member of the OmniScriptum S.R.L Publishing group
str. A.Russo 15, of. 61, Chisinau-2068, Republic of Moldova Europe
Printed at: see last page
ISBN: 978-620-4-17075-6

GENERAL INDEX

GENERAL TABLE OF CONTENTS i
SUMMARY 1
INTRODUCTION 2
CHAPTER I 4
RESEARCH PROBLEM 4
1.1 PROBLEM STATEMENT 4
FORMULATION OF THE PROBLEM 7
1.2.1 GENERAL QUESTION: 7
1.2.2. SPECIFIC QUESTIONS 7
1.3. JUSTIFICATION OF THE INVESTIGATION 7
1.4. RESEARCH OBJECTIVES 9
1.4.1. General objective 9
1.4.2. Specific objectives 9
CHAPTER II 11
THEORETICAL FRAMEWORK 11
2.1 BACKGROUND 11
2.2. FRAME OF REFERENCE 15
2.2.1. TOTAL FIBRE 15
2.2.1.1. FIBRE PROPERTIES 17
2.2.1.2. QUINOA, KIWICHA AND CAÑIHUA FIBRE 19
2.2.1.3. COMPONENTS OF THE SOLUBLE FIBRE OF QUINOA, KIWICHA AND CAÑIHUA 20
2.2.1.4 INSOLUBLE FIBER COMPONENTS OF QUINOA, KIWICHA AND CAÑIHUA 20
2.2.1.5 FIBRE INTAKE RECOMMENDATIONS 21
2.2.2 YOGHURT 22
2.2.2.1 TYPES OF YOGHURT 22
2.2.2.2 BENEFITS OF YOGURT 23
2.2.2.3 ELABORATION PROCESS 23
2.2.2.4 PHYSICOCHEMICAL AND MICROBIOLOGICAL CHARACTERISTICS OF YOGHURT 27
2.2.3 SENSORY ANALYSIS 27
2.2.4 SHELF LIFE 30

2.2.4.1 METHODOLOGIES FOR DETERMINING SHELF LIFE IN FOODSTUFFS 30
2.2.4.4 ACCELERATED SHELF LIFE TESTING (ASLT) 34
CHAPTER III 37
METHODOLOGY 37
3.1 SCOPE OR PLACE OF STUDY 37
3.2 POPULATION AND SAMPLE 37
3.2.1 POPULATION 37
3.3 METHODS 39
3.3.1 TO IDENTIFY THE PHYSICOCHEMICAL AND MICROBIOLOGICAL CHARACTERISTICS OF THE MOST ACCEPTABLE YOGHURT WITH DIFFERENT LEVELS OF FIBRE ISOLATED FROM QUINOA, KIWICHA AND CAÑIHUA BRAN CONSIDERING THE ACCEPTABLE PARAMETERS 39
3.3.1.1 METHOD FOR OBTAINING THE GRAINS 39
3.3.1.2 OBTAINING BRAN 39
3.3.1.3 FIBRE EXTRACTION (AOAC 2000) 40
3.3.1.4 DETERMINATION OF THE SENSORY ANALYSIS (ANDALZUA METHOD 1994) 42
3.3.1.5 pH DETERMINATION (AOAC 2002) 42
3.3.1.6 DETERMINATION OF ACIDITY (AOAC 2002) 43
3.3.1.7 FAT DETERMINATION (AOAC 2002) 43
3.3.1.8 DETERMINATION OF MICROBIOLOGICAL ANALYSIS (MORENO 2000) 44
3.3.2 TO ESTABLISH THE SHELF LIFE OF THE MOST ACCEPTABLE YOGHURT WITH FIBRE ISOLATED FROM QUINOA, KIWICHA AND CAÑIHUA BRAN 47
3.3.2.1 DIRECT METHOD: Recommended Steps 47
3.3.2.2 ACCELERATED TESTING: 47
3.3.3 TO DETERMINE THE COMPOSITION OF THE MOST ACCEPTABLE YOGHURT WITH FIBRE ISOLATED FROM QUINOA, KIWICHA AND CAÑIHUA BRAN 48
3.4 STATISTICAL ANALYSIS 48
3.4.1 TO DETERMINE THE SHELF-LIFE EFFECTS OF THE MOST ACCEPTABLE YOGHURT WITH FIBRE ISOLATED FROM QUINOA, KIWICHA AND CAÑIHUA BRAN 48
3.4.2 TO ESTABLISH THE SHELF LIFE OF THE MOST ACCEPTABLE YOGHURT WITH FIBRE ISOLATED FROM QUINOA, KIWICHA AND CAÑIHUA BRAN 49
3.4.3 TO DETERMINE THE COMPOSITION OF YOGHURT WITH FIBRE ISOLATED FROM QUINOA, KIWICHA AND CAÑIHUA BRAN 50
4.1 IDENTIFICATION OF THE PHYSICOCHEMICAL AND MICROBIOLOGICAL CHARACTERISTICS OF THE MOST ACCEPTABLE YOGHURT WITH DIFFERENT

LEVELS OF FIBRE ISOLATED FROM QUINOA, KIWICHA AND CAÑIHUA BRAN. 54

4.2 SHELF LIFE OF THE MOST ACCEPTABLE YOGHURT WITH FIBRE ISOLATED FROM QUINOA, KIWICHA AND CAÑIHUA BRAN. 57

4.3 COMPOSITION OF THE MOST ACCEPTABLE YOGHURT WITH DIFFERENT LEVELS OF FIBRE ISOLATED FROM QUINOA, KIWICHA AND CAÑIHUA BRAN. 64

CONCLUSIONS 66

RECOMMENDATIONS 67

BIBLIOGRAPHY 68

ANNEXES 73

SUMMARY

The present research work was carried out with the objectives of Identifying the physicochemical and microbiological characteristics of the most acceptable yogurt with different levels of fiber isolated from quinoa, kiwicha and cañihua bran considering the established parameters. To establish the shelf life of the most acceptable yogurt with fiber according to the levels of fiber isolated from quinoa, kiwicha and cañihua bran. To determine the composition of the most acceptable yogurt with different levels of fiber from quinoa, kiwicha and cañihua bran during shelf life. To obtain the bran, the grains were subjected to the process of cleaning, de-stoning, drying, milling and sieving, in order to improve the bran extraction yield. From the bran fiber was obtained by the AOAC method (2002), the same that was added in proportion of 3%, 4% and 5% to yogurt, to evaluate it by the method of hedonic test and thus obtain the most acceptable yogurt, the same that was evaluated according to their physicochemical and microbiological characteristics, shelf life was evaluated by the direct method and its composition: moisture, protein, fat, carbohydrate, fiber, total solids and ash making use of the AOAC methods and for the microbiological characteristics through total count of mesophilic aerobic microorganisms, total coliforms, molds and yeasts. Yogurt with fiber isolated from quinoa bran at 5%, yogurt with fiber isolated from kiwicha bran at 5% and yogurt with fiber isolated from cañihua bran at 5% were the most acceptable. In relation to the physicochemical characteristics, the yoghurt with isolated fibre from quinoa bran showed the highest values for pH, acidity, protein and fat; and the yoghurt with isolated fibre from kiwicha bran showed the highest value for total fibre; as for the microbiological characteristics, all the samples under study showed no coliforms or moulds. In relation to the shelf life for pH and acidity during the 35 days that lasted the evaluation are within the established parameters. As for the composition: the most acceptable yogurt with isolated fiber from quinoa bran has the highest proportion of moisture, protein, fat and ash; the yogurt with isolated fiber from kiwicha bran has a higher proportion of carbohydrates and total solids and the one with isolated fiber from cañihua bran has a higher proportion of fiber.

Keywords: sensory analysis, physicochemical characteristics, bran, yoghurt, yoghurt

INTRODUCTION

Yoghurt, a product made from heat-treated milk inoculated with lactic acid bacteria, *Lactobacillus bulgaricus* and *Streptococcus thermophilus, is a* product that comes in different forms, such as plain and whipped. The physicochemical properties and texture are influenced by several factors, such as the composition and formulation of the milk, the heat treatment given to the milk, the combination of lactic acid bacteria used, the rate of acidification of the milk and the storage time.

Yogurt, besides being an excellent food of dairy origin, has been used as a suitable medium for the incorporation of other nutrients such as food components, prebiotic ingredients, fiber and calcium, among others.

As for its formulation, isolated fiber has been incorporated, a very useful component that prevents constipation, reduces the risk of colon cancer, as well as reduces cholesterol levels in the blood. Fiber, previously considered as a non-nutritive component, has taken great nutritional importance; it is considered a nutrient and its deficiency can have adverse consequences on health, as it has been related to the incidence of certain chronic diseases.

Fiber is a very important food component that consists of plant material resistant to the endogenous enzymes of the digestive tract of mammals. That is why, the elaboration of a functional yogurt based on fiber, and with controlled fat level, is a very good option to complement the daily diet, since apart from being a delicious, nutritious, healthy, and convenient food, it is also novel for the consumer.

The incorporation of new ingredients in the yogurt formulation changes the original structure of the gel both physically and chemically, so it is important to know its effects.

CHAPTER I

RESEARCH PROBLEM

1.1 PROBLEM STATEMENT

From the technological point of view, the fiber is lost during the cooking processes by expansion of quinoa and kiwicha, not giving them the respective utility, that is to say, to incorporate them into products such as yogurt, thus promoting the consumption of the fiber of these grains.

The population is not in the habit of consuming frequent sources of fiber (the recommended 25 to 30 g per day), which has health consequences. An insufficient intake of dietary fibre contributes to a cluster of chronic disorders such as constipation, diverticulitis, diabetes, obesity, coronary heart disease and cancer of the large intestine. These hypotheses have developed mainly from early observational studies. All of these disorders have a multifactorial etiology and over time, experimental research has contributed to establishing knowledge of their relationships with fiber (Rubio, 2002).

Dietary fiber is known to promote beneficial physiological effects such as laxative, and/or attenuates blood cholesterol levels, as well as attenuates blood glucose.

Constipation is relieved and prevented by dietary fibre intake, increases in faecal bulk and weight are important, but are not the only factors involved (Ilsi, 2006). Wheat bran has been shown to be the most effective in increasing stool bulk, although isolated cellulose is also effective and increases stool bulk to a greater degree than isolated fermentable fibres such as pectin. However, all carbohydrates that are not absorbed can increase laxation through water uptake, osmotic effects of degradation products and increased bacterial mass. On the other hand, when faced with diverticulosis of the colon, the intake of dietary fibre protects against the disorder and relieves the symptoms. Non-viscous fibres such as cellulose are particularly effective in this respect, as are cereal bran foods containing fibre.

These protective effects may involve increased stool weight, decreased transit time, and decreased intracolonic pressure.

Also, the apparent protective effect of dietary fiber on the cardiovascular system includes changes in cholesterol absorption and bile acid reabsorption, alterations in lipoprotein production in the liver, and changes in lipoprotein clearance from the bloodstream. All of this can result in lower plasma total cholesterol and LDL (Low Density Lipoprotein) levels, which would decrease the risk of coronary heart disease. Fiber slows the absorption of fat and carbohydrates from the small intestine and has concomitant effects on insulin metabolism. It may also reduce the level of circulating triglycerides and as a result, reduce the risk of coronary heart disease (Ilsi, 2006).

Increased consumption of highly viscous fibres such as ß-glucans, pectins and guar gum are associated with significant reductions in blood cholesterol levels in overweight and obese individuals, as well as in subjects with hyperlipidaemia. However, dietary fibre components such as non-viscous fibres (e.g. wheat fibre and cellulose) do not influence blood lipids (Kin, 2000).

Various human intervention research has shown that isolated viscous fibres (ß-glucans, oat bran, pectins, guar gum) have cholesterol-lowering properties, but only if their intake is much higher than the levels consumed in typical diets (ILSI, 2006).

Among the components of the Andean cereal varieties studied in whole grains (quinoa, kiwicha and cañihua), their total dietary fiber (TDF) content stands out, which varied between 9.29 and 20.04 percent, a component made up of insoluble dietary fiber (IDF) and soluble dietary fiber (SDF). (Ligarda et al. 2012).

For the acidity of the sample containing yacon, the initial and final value was 0.87 and 0.91% expressed in lactic acid, respectively. The control yogurt presented initial and final value 0.63 and 0.88% of lactic acid, respectively, this is due to what is mentioned (Apolinario, 2014) who showed that the addition of inulin significantly reduces the fermentation time (incubation), increasing the growth of biomass and increasing the levels of lactic acid in yogurt. This may be due to the fact that inulin and fructo oligosaccharides (FOS) are more easily assimilated substrates by lactic acid bacteria (LAB) during the incubation and storage period. It can also be attributed to the fact that during storage under refrigerated conditions, microbial activity by the lactic acid bacteria present in the yoghurt occurred, as reported by Lubbers et al. (2004), in yoghurt stored for more than 20 days under refrigeration.

The Andean region is one of the great centers of origin and domestication of numerous crops, such as quinoa (*Chenopodium quinoa* Willd), kiwicha (*Amaranthus caudatus* L.), cañihua (*Chenopodium pallidicaule* Aellen.) that are being studied through research such as the present one.

In view of this approach, the following questions arise:

FORMULATION OF THE PROBLEM

1.2.1 GENERAL QUESTION:

Does yoghurt made with different levels of fibre isolated from the bran of quinoa (*Chenopodium quinoa* Willd), kiwicha (*Amaranthus caudatus* L) and cañihua (*Chenopodium pallidicaule* Aellen) have acceptable physicochemical, microbiological and sensory characteristics?

1.2.2. SPECIFIC QUESTIONS

Are the physicochemical and microbiological characteristics of the most acceptable yoghurt with different levels of fibre isolated from quinoa, kiwicha and cañihua bran within acceptable parameters?

Will the composition of different levels of fiber in the most acceptable yogurt enriched with fiber isolated from quinoa, kiwicha and cañihua bran be altered during shelf life?

What is the shelf life of the most acceptable yogurt with isolated fiber from quinoa, kiwicha and cañihua?

1.3. JUSTIFICATION OF THE INVESTIGATION

Fiber includes polysaccharides, oligosaccharides, lignin and associated plant substances. Dietary fibers promote physiological beneficial effects such as laxative, and/or attenuates blood cholesterol levels and/or attenuates blood glucose. Through the present research, fiber will be isolated from quinoa, kiwicha and cañihua bran.

Fermented milk is a dairy product obtained through fermentation of milk by the addition of bacteria that acidify it and are responsible for metabolic transformations in carbohydrates, proteins and lipids, which lead to the development of its characteristic flavour and texture. The most important transformation is lactic fermentation, which uses lactose from milk as a substrate. Mainly glucose from the hydrolysis of lactose gives rise to lactic acid and small amounts of a number of compounds that contribute to the aroma. As a consequence of the lowering of the pH, the development of undesirable microorganisms is hindered, the colloidal calcium and phosphorus in the milk pass into the soluble form and the major proteins, the caseins, free of calcium precipitate in the form of a fine coagulum, which facilitates the action of human proteolytic enzymes and consequently favours digestibility. One of the best known types of fermented milks is yoghurt. Only two types of bacteria are in charge of the fermentation to obtain yoghurt, *Streptococcus thermophilus* and *Lactobacillus delbrueckii subsp.* Thanks to these bacteria, which are active in the final product, the lactose in the milk is transformed into lactic acid. There are other types of fermented milks with similar physical-chemical composition obtained with other microbial species and, more recently, a new generation of fermented milks has been developed, whose main characteristic is that other live microorganisms with other probiotic effects are associated with the usual lactic ferments. Mainly strains of the *Lactobacillus* and *Bifidobacterium* genera are used.

The consumption of milk and other dairy products should be considered in the context of a varied and balanced diet, as the consumption of some foods is associated with healthier behaviours and vice versa. In different studies it has been observed that a high intake of dairy products and in particular the intake of yoghurt, is accompanied by a higher quality of the diet. Through the present investigation it was demonstrated that yogurt with fiber isolated from quinoa bran, kiwicha and cañihua have acceptable physicochemical and microbiological characteristics, with a shelf life within the established parameters.

The present research work does not have negative environmental effects or consequences because for the elaboration of yogurt raw material (whole milk) was used in its entirety, not generating other products such as whey, which would need treatment to not contribute to environmental pollution, also the bran from which the fiber was isolated was used not generating waste to be eliminated.

1.4. RESEARCH OBJECTIVES

1.4.1. General objective

To determine the physicochemical, microbiological and sensory characteristics of the most acceptable yoghurt with different levels of fibre isolated from the bran of quinoa (*Chenopodium quinoa* Willd.), kiwicha (*Amaranthus caudatus*) and cañihua (*Chenopodium pallidicaule* Aellen).

1.4.2. Specific objectives

a) To identify the physicochemical and microbiological characteristics of the most acceptable yogurt with different levels of fiber isolated from quinoa, kiwicha and cañihua bran considering the acceptable parameters.

b) Establish the shelf life of the most acceptable yoghurt with fibre according to the levels of fibre isolated from quinoa, kiwicha and cañihua bran.

c) To determine the most acceptable yogurt composition with different levels of fiber isolated from quinoa, kiwicha and cañihua bran during shelf life.

CHAPTER II

THEORETICAL FRAMEWORK

2.1 BACKGROUND

Ruíz et al. (2009), elaborated firm yogurt with the incorporation of probiotic strains (*Bifidobacterium spp.* and *Lactobacillus acidophilus*) and inulin using three yogurt formulations: F1 = Pasteurized milk + Lactic Culture + Probiotic Strains, F2 = Pasteurized milk + Lactic Culture + Probiotic Strains + Inulin, F3 = Pasteurized milk + Lactic Culture (Control). The raw milk complied with the requirements established in the COVENIN standard (903:93). Yogurts complied with the microbiological requirements of COVENIN (2392:01). Formulation 2 (probiotics and inulin), showed greater physicochemical stability during storage time and also did not present the phenomenon of syneresis.

Rodriguez et al. (2002) developed a process for the elaboration of lactose-free yogurt from goat milk, performing lactose hydrolysis and fermentation simultaneously. Three doses of an acid β-galactosidase of fungal origin (*Aspergillus oryzae*) were used: 1253, 2506 and 3759 units/L, using a sample without enzyme as a reference. After incubation for 5 hours at 38 °C and subsequent storage for 24 hours at 4 °C, the concentrations of glucose, galactose, lactose, pH and titratable acidity were determined. Considering a lactose concentration in milk of 4.42 g. 100 mL^{-1}, the highest percentage of hydrolysis obtained in yogurt (82.6%, $p < 0.0001$) was achieved with 2506 units/L of enzyme, while without the addition of enzyme it was 48.5%. It is concluded that the process used would allow, in only one stage, the elaboration of a yogurt that besides having the hypoallergenic and nutritional properties of goat milk, due to its low content in lactose, it would become a food with functional characteristics of great utility and prominence for those people who suffer intolerance to milk sugar.

Rojas et al. (2007) determined the presence of organochlorine insecticide residues (IOC) in yoghurt of three commercial brands and different storage times, using the liquid-liquid extraction technique recommended by the Official Association of Analytical Chemists AOAC (1997). Three commercial brands of firm natural yogurt, produced in Venezuela and distributed in the state of Zulia, called A, B and C, and three storage times (0, 15 and 30 days prior to the expiration date) were selected at a temperature of 4°C, obtaining a total of 54 samples. Of the samples analyzed, 48 (88.9%) showed IOC residues. In the 3 brands, endosulfan was found in the highest concentration, with values of 0.1137 mg kg^{-1} of fat for brand A; 0.1105 mg kg^{-1} of fat for brand B and 0.1927 mg kg^{-1} of fat for brand C, while aldrin (0.0037 mg kg^{-1} of fat for brand A), endrin (0.0076 mg kg^{-1} of fat for brand B) and lindane (0.0045 mg kg^{-1} of fat for brand C) were detected in lower concentrations. Endrin and mirex were not detected for brand A; aldrin, methoxychlor and mirex were not detected for brand B and endrin, DDT and mirex were not detected for brand C. In some of the samples, the concentration of the IOCs exceeded the maximum residue limits (MRLs); with the exception of hexachlorobenzene for all 3 brands; heptachlor, aldrin, dieldrin and DDT for brand A; p,p'-DDE for brand B and lindane for brand C. No significant differences were found in the concentration of IOCs by effect of storage time at 4°C. There was no decrease in the IOC concentration during storage.

Monitoring of IOC residues in food produced in Venezuela is recommended Revista Científica.

Acevedo et al. (2007) evaluated the effect of different proportions of goat milk (c) and cow milk (v) (0%c/100%v, 30%c/70%v, 50%c/50%v, 70%c/30%v and 100%c/0%v), on the pH, viscosity and syneresis of a strawberry whipped yoghurt, during days 1, 7, 14 and 21 of refrigerated storage at 4-5°c. The pH decreased in storage sharply in the first seven days and indistinctly for all formulations ($p \leq 0.05$) from initial ranges of 4.35-4.40 to 4.25-4.30. During the first seven days the viscosity of all samples increased and then decreased until day 21. Samples with 100% goat's milk had lower viscosity ($p \leq 0.05$) (mean = 11277 cp) than those made with 100% cow's milk (mean = 19979 cp). The syneresis for all samples decreased with time. The sample with the highest syneresis during the entire period was 100% cow's milk (mean = 9.4%), while the lowest was 100% goat's milk (mean = 2.1%). For syneresis, a significant interaction ($p \leq 0.05$) was found between day of storage and type of milk, concluding that syneresis decreased with time and with higher cow's milk content. The effect of different formulations (30%c/70%v, 50%c/50%v, 70%c/30%v and 100%c/0%v), on overall liking, as well as color and texture acceptance, was evaluated with 105 judges. The formulation with the highest overall liking ($p \leq 0.05$) was the 30% goat milk formulation, which on average reached a value of 8.1 on a 10 cm hybrid hedonic scale.

Miranda et al. (2014) elaborated a fermented beverage from cheese whey at the Combinado Lácteo "La Hacienda" in the city of Bayamo (Granma, Cuba). Five experimental runs of 200 liters were carried out at pilot plant scale with each of the experimental variants (Variant 1: potassium sorbate: 0.0% vs. Variant 2: potassium sorbate: 0.03%) to establish the main physicochemical, sensory, nutritional, microbiological and durability characteristics of the fermented beverage 24 hours after being inoculated. The titratable acidity curve was determined by simple regression analysis using the equation Y = a + b*x (Y = acidity; x = time). A significant correlation ($r = 0.95$; $p < 0.05$) was found between product acidity and fermentation time. The physicochemical, microbiological and sensory indicators obtained guarantee a product of good quality and innocuous. A titratable acidity as lactic acid of 0.63%, a total solids content of 19.43%, and a viscosity of 26 seconds were obtained. The nutritional composition of the fermented beverage was as follows: Crude protein: 1.22%; Carbohydrate: 17.53%; Dietary energy: 77.52 Kcal; respectively. The viable lactic acid bacteria count was 1.2×10^7 cfu.mL^{-1}. Consumer tests determined a mean score of 6 (corresponding to "I like it very much"). The use of potassium sorbate as a preservative prolonged the shelf life of the beverage from 7 days to 28 days.

Bueno-Solano et al. (2009) determined riboflavin content in commercial dairy products, milk for direct consumption and yoghurt, by high performance liquid chromatography (HPLC) with fluorescence detection. The riboflavin content of yoghurt samples ranged from 0.289 - 3.078 µ/g BS, while in ultra-pasteurized and pasteurized milk from 0.61-13.64 µg/g BS and 11.73-15.41 µg/ g BS, respectively. Milks for direct consumption and commercial yoghurt are sources of riboflavin, and regular consumption may help to meet daily requirements.

Tola (2006) elaborated yogurt with raw material (milk) concluding that: the values found of the shelf life of yogurt in its different characteristics (with preservative, without preservative and natural), establish an inverse relation between the conservation temperature and the shelf life (the higher the temperature the shorter the shelf life), on the other hand, the influence of the use of preservative is determinant to extend the shelf life of yogurt. Finally, the non-addition of preservative involves a rapid rise in acidity between the end of the preparation and the first day of control.

Ligarda et al. (2012) evaluated the dietary fiber content and its components in 3 varieties of quinoa (*Chenopodium quinoa* Willd), 3 varieties of kiwicha (*Amaranthus caudatus* L.) and 3 varieties of cañihua (*Chenopodium pallidicaule* Aellen). The following varieties of each Andean grain were chosen for their high content of soluble dietary fiber (SDF): quinoa: Salcedo INIA variety, cañihua: Cupi variety and kiwicha: Morocho variety. The study concludes that: among the components of the Andean cereal varieties studied in whole grains (quinoa, kiwicha and cañihua), their total dietary fiber (TDF) content stands out, which varied between 9.29 and 20.04 percent, a component made up of insoluble dietary fiber (IDF) and soluble dietary fiber (SDF). The quinoa varieties Salcedo INIA, kiwicha Morocho and cañihua Cupi had the highest contents of SDF in whole grain (4.68, 2.13 and 3.79 g/100 g), respectively, compared to the other varieties of each Andean grain studied.

2.2. FRAME OF REFERENCE

2.2.1. TOTAL FIBRE

Mataix (2008) indicates that the term fibre is applied to those substances of vegetable origin, mostly carbohydrates, not digested by human enzymes and with the peculiarity of being partially fermented by colonic bacteria. Insoluble fibre includes cellulose, hemicelluloses and lignin. As functional actions are attributed to it: the increase of the fecal bolus and the stimulation of intestinal motility; the greater need of chewing, relevant in modern societies victims of compulsive ingestion and obesity, the increase of bile acid excretion and antioxidant and hypocholesterolemic properties.

Valenzuela and Maiz (2006) point out that, in recent times, fibre has been classified, according to its degree of solubility, into soluble and insoluble. In general, it is accepted that soluble fiber is viscous and fermentable, while insoluble fiber is not viscous and is poorly fermentable. This is questionable, since inulin and FOS (Fructo-oligosaccharides) are soluble and fermentable, but have a very low viscosity.

Mataix (2008) indicates that soluble fibre is mainly represented by pectins, gums, mucilages and some hemicelluloses; its main characteristic is its ability to trap water and form viscous gels, which determines its laxative power. Likewise, by significantly increasing the quantity and consistency of the faecal bolus, a positive effect is achieved in the case of diarrhoea. In addition, it slows down the digestive process, the transit and absorption of carbohydrates, as well as an additional feeling of fullness. Like insoluble fibre, it reduces the absorption of bile acids and has cholesterol-lowering activity. As for lipid metabolism, it seems to lower triglyceride levels, LDL cholesterol (low density lipoproteins) and reduce postprandial insulinemia. A fundamental characteristic of soluble fiber is its ability to be metabolized by colonic bacteria, with the consequent production of gases (flatulence, fecal propulsion) and short-chain fatty acids: acetate, propionate and butyrate. The first two can be absorbed and used for energy. Propionate has an inhibitory action on hydroxymethylglutaryl-coenzyme A reductase, a limiting step in the synthesis of endogenous cholesterol.

2.2.1.1. FIBRE PROPERTIES

García (2000) refers that it is considered that terms such as soluble/insoluble, fermentable/non-fermentable and viscous/non-viscous should disappear from the nomenclature on fiber, these properties are the basis of their physiological benefits, so from a practical point of view it would be an appropriate classification, deriving widely accepted concepts such as: fermentable, soluble and viscous fiber and scarcely fermentable, insoluble and non-viscous fibers.

These properties depend on the composition of the specific fibre we are administering, not on the fibre in general.

Kin (2000) states that soluble fibers in contact with water form a reticulum where it is trapped, originating solutions of high viscosity. The effects derived from the viscosity of the fibre are responsible for its actions on lipid and hydrocarbon metabolism and, in part, its anticarcinogenic potential. Insoluble or poorly soluble fibres are able to retain water in their structural matrix forming low viscosity mixtures; this produces an increase in faecal mass which accelerates intestinal transit. This is the basis for using insoluble fibre in the treatment and prevention of chronic constipation. On the other hand, it also contributes to decrease the concentration and contact time of potential carcinogens with the colon mucosa.

Mataix (2008) indicates that the particle size of the fibre can influence its capacity to capture water; the processing of the feed, such as cereal milling, and chewing will be influential factors.

It is also interesting to note that water retention is also affected by the fermentation processes that dietary fibre can undergo in the large intestine.

Sastre (2003) states that in the colon there are basically two types of fermentation: saccharolytic fermentation and proteolytic fermentation.

The main products of fibre fermentation are short-chain fatty acids (SCFA), gases (hydrogen, carbon dioxide and methane) and energy.

Glucose polymers are hydrolyzed to monomers by the action of extracellular enzymes of colon bacteria. Metabolism continues in the bacteria until pyruvate is obtained from glucose in the Embdem-Meyerhoff metabolic pathway. This pyruvate is converted into short chain fatty acids (SCFA): acetate, propionate and butyrate, in an almost constant molar ratio 60:25:15. In smaller proportions, valerate, hexanoate, isobutyrate and isovalerate are also produced. It can be calculated for example that 64.5 moles of fermented carbohydrates produce 48 moles of acetate, 11 moles of propionate and 5 moles of butyrate.

Gibson (2004) states that proteolytic fermentation produces nitrogenous derivatives such as amines, ammonium and phenolic compounds, some of which are carcinogenic.

More than 50 percent of the fiber consumed is degraded in the colon, the rest is eliminated in the stool.

All types of fibre, with the exception of lignin, can be fermented by intestinal bacteria, although in general the soluble ones are fermented in greater quantities than the insoluble ones. Cellulose has a fermentation capacity between 20 and 80%; hemicellulose from 60 to 90%; guar fibre, resistant starch and fructo-oligosaccharides have a capacity of 100%. Wheat bran only 50%.

On the other hand, the fibre itself, the gases and the SCFA generated during fermentation are capable of stimulating the growth of the number of microorganisms in the colon. It is estimated that the regular intake of 20 grams/day of guar gum (highly fermentable) would increase the weight of the stool by 20%, with the advantage of the mass and anticarcinogenic effect that this entails.

Ingestion of fructooligosaccharides (functional fibre) can increase the numerical representation of bifidobacteria tenfold, in what has been termed the prebiotic effect: "non-digestible components of the diet that are beneficial to the host because they produce the selective growth and/or activity and/or of one or a limited number of bacteria in the colon.

2.2.1.2. QUINOA, KIWICHA AND CAÑIHUA FIBRE

Ligarda et al. (2012) indicate the FDT values in quinoa, kiwicha and cañihua. Cañihua presented high TDF contents followed by kiwicha and quinoa which have similar contents. Cañihua grains have high FDI content followed by kiwicha and quinoa grains which are low in FDI and high in SDF. Cañihua grains have higher SDS content compared to kiwicha grains as shown in the table below:

TABLE 1
RANGE OF DIETARY FIBER CONTENT IN QUINOA, KIWICHA AND CAÑIHUA (g/100g DRIED MATTER) EVALUATED

Andean grain	Soluble fiber (%)	Insoluble fiber (%)	Total dietary fibre (%)
Quinoa	3,2 – 5,3	6,1 – 7,4	10,4 – 11,5
Kiwicha	1,9 – 2,4	8,5 – 9,3	10,9 – 11,3
Cañihua	2,3 – 4,1	15,6 – 18,7	18,7 – 21,9

Source: Ligarda et al. (2012)

2.2.1.3. COMPONENTS OF THE SOLUBLE FIBRE OF QUINOA, KIWICHA AND CAÑIHUA

Ligarda et al (2012) indicate that the content of β-glucans in Andean grains is very low. The kiwicha crops have higher contents than quinoa and cañihua). The pentosan content was high in kiwicha grains and lower in quinoa and cañihua grains, as shown in the following table:

TABLE 2
RANGE OF FIBER CONTENT IN QUINOA, KIWICHA AND CAÑIHUA (g/100g DRIED MATTER) EVALUATED

Andean grain	Pentosans (%)	β-glucans(%)
Quinoa	1,8 – 2,0	0,09 – 0,10
Kiwicha	0,8 – 1,1	0,63 – 0,97
Cañihua	0,2 – 1,1	0,04 – 0,11

Source: Ligarda et al (2012)

2.2.1.4 INSOLUBLE FIBRE COMPONENTS OF QUINOA, KIWICHA AND CAÑIHUA

Ligarda et al. (2012) indicate that cellulose, Klason lignin and resistant starch were high in cañihua followed by quinoa and kiwicha. Quinoa and kiwicha grains have similar cellulose and Klason lignin contents.

Andean grains, in general, have low contents of resistant starch. Klason lignin (obtained by solubilizing cell wall polysaccharides with H_2SO_4) has physicochemical properties different from soluble lignins such as dioxane hydrochloric lignin, as shown in the table below:

TABLE 3

RANGE OF FIBER CONTENT (RESISTANT SULFIDE, CELLULOSE AND LIGNIN) IN QUINUA, KIWICHA AND CAÑIHUA (g/100g OF DRY MATTER) EVALUATED

Andean grain	Resistant starch (%)	Cellulose (%)	Lignin Klason (%)
Quinoa	0,20 – 0,33	3,0 – 4,4	2,7 – 4,3
Kiwicha	0,10 – 0,12	3,3 – 4,5	3,3 – 4,4
Cañihua	0,24 – 0,34	6,2 – 9,6	6,0 – 9,6

Source: Ligarda et al. (2012)

2.2.1.5 FIBRE INTAKE RECOMMENDATIONS

Olivares et al. (1994) indicates that for adults an intake of 20-35 g/day or approximately 10-14 g of dietary fibre per 1,000 kcal is suggested.

In general, the fiber consumed should have a ratio of 3/1 between insoluble and soluble.

2.2.2 YOGHURT

Romero (2008), states that yoghurt is understood as the coagulated milk product obtained by lactic fermentation through the action of *Lactobacillus bulgaricus* and *Streptococcus thermophillus* from pasteurized milk, partially or totally or partially skimmed pasteurized milk, with or without the addition of pasteurized cream, whole, semi-skimmed or skimmed milk powder, whey powder, milk proteins and/or other products from the fractionation of milk.

Yoghurt results from the development of two bacteria, *Lactobacillus bulgaricus and Streptococus thermophillus. Lactobacillus bulgaricus* is a lactic bacteria that optically develops between 42 and 45°C, strongly acidifying the medium.

Streptococcus thermophillus multiplies well between 37 to 42°C and performs the function of flavoring yogurt, so it is advisable to incubate the yogurt at 42°C, so that a 50:50 ratio of each type of bacteria is maintained.

2.2.2.1 TYPES OF YOGHURT

Romero (2008) indicates that there are the following types of yogurt:

- ✓ Sweetened yoghurt is natural yoghurt to which authorised sweeteners have been added.
- ✓ Yoghurt with fruit, juices and/or other natural products is natural yoghurt to which fruit, juices and/or other natural products have been added.
- ✓ Flavoured yoghurt, natural yoghurt to which authorised flavouring agents have been added.
- ✓ Natural Yogurt
- ✓ Sweetened yoghurt is plain yoghurt to which sugar or edible sugars have been added.

2.2.2.2 BENEFITS OF YOGHURT

Garcia (2000), points out that from a nutritional and health point of view, fermented milks provide additional nutrients to those of the fresh product, such as B vitamins and a greater amount of protein in concentrated products such as yogurt. In addition, proteins have a higher biological value due to the prehydrolysis they undergo by proteases produced by lactic acid bacteria. Fat and lactose are also more digestible in these products than in milk, due to the action of microbial enzymes.

Fermented milks are convenient foods for people suffering from lactose intolerance, since this problem does not occur when these foods are consumed; the probable explanation is the presence of microbial lactases in the intestinal tract.

2.2.2.3 ELABORATION PROCESS

The following is the process of making yoghurt.

FIGURE 1
YOGURT ELABORATION PROCESS

RECEPCIÓN DE LA LECHE
ESTANDARIZACIÓN DE LA LECHE
ADICIÓN DE COMPONENTES MINORITARIOS
DESAIREADO
HOMOGENIZACIÓN DE LA LECHE
PASTERIZACIÓN
REFRIGERACIÓN
ADICIÓN DE FERMENTOS

ADICIÓN DE COMPONENTES MINORITARIOS
ENVASADO Y TAPADO
FERMENTACIÓN
REFRIGERACIÓN Y ALMACENADO
YOGUR FIRME

FERMENTACIÓN
REFRIGERACIÓN
ADICIÓN DE COMPONENTES MINORITARIOS
BATIDO
ENVASADO Y TAPADO
REFRIGERACIÓN Y ALMACENADO
YOGUR BATIDO

Source: Ligarda et al. (2012)

Hernandez (2003) details that the yogurt elaboration process is as follows:

Raw Material: Yogurt is made with both whole and skim milk, preferably cow's milk, although in other countries goat's, mare's or buffalo's milk is used. Reconstituted powdered milk may also be used. The milk must be free of antibiotics, because their presence inhibits the development of the microorganisms that carry out the fermentation.

Standardization: To increase the total solids content of milk, it is first necessary to standardize the amount of fat. The fat content of yoghurt should be between 0.5% for nonfat and 3.5% for whole yoghurt.

Homogenization: This stage is generally carried out before pasteurization, but can be done afterwards. It consists of subjecting the milk to high pressures (between 2.6 and 6.8 KPa) in order to reduce the size of fat droplets and other constituents so that they are better dispersed. The result is a more viscous, more stable yoghurt with better organoleptic characteristics.

Pasteurization: This is one of the most important stages of this process because:

✓ Most of the flora contained in the milk is eliminated. The reduction of the flora associated with the milk allows the growth of microorganisms (yoghurt producers) free of competition, with all the nutrients of the milk at their disposal.

✓ The inactivation of enzymes that affect the organoleptic characteristics of yogurt is achieved.

✓ The milk proteins are denatured. By denaturing the proteins, peptides are released that contribute to the growth of the inoculated microorganisms.

Post-pasteurization cooling: Milk should be cooled to the temperature necessary for optimal growth of microorganisms, which ranges between 40 and 45°C. Cooling can be carried out in two ways:

✓ The milk is passed through a plate heat exchanger.

✓ In the same pasteurization tank, cold (instead of hot) water is passed through the reactor jacket.

Inoculation and fermentation: The starter culture is composed of the microorganisms *S. thermophilus and L. bulgaricus* in a 1:1 ratio, which guarantees an adequate yogurt consistency and a pleasant aroma. The starter culture is inoculated in a proportion between 1 and 5% of the initial amount of milk used. It must be mixed very well with the milk to ensure adequate distribution of the microorganisms. At this point the fermentation process begins. Fermentation takes an average of three to six hours, at a temperature between 40 and 45°C (104 and 113°F). The fermentation time depends on the incubation temperature and the lactic acid production capacity of the microorganisms. The process should be stopped when a lactic acid concentration between 0.70 and 1.1% w/v is reached. In this acid concentration range, the pH value is between 4.6 and 3.7.

Post-fermentation cooling: When the desired acidity is reached, the fermentation process must be stopped. To stop fermentation, the temperature is lowered, because the microorganisms involved in the process are not able to grow at temperatures below 10°C; furthermore, at low temperatures, the activity of the enzymes generated by the microorganisms is suspended. The recommended temperature is cooling (5°C). At the same time, cooling has a positive effect, as it increases the firmness of the gel.

Agitation and the addition of fruits: The addition of fruits to yogurt gives it greater consumer acceptance. Once the yogurt is cold, it must be stirred carefully to break the clot or gel; if the agitation is done abruptly, the gel loses its viscosity. During this stage the fruit is added, previously prepared in the form of pieces of puree, in percentages that vary from 5 to 25% of the final product. The fruits must be previously heat-treated, otherwise they are a source of fungi and yeasts that will contaminate the yogurt and reduce its shelf life.

Packaging: When the yoghurt has cooled and the fruits have been added, the product is packaged. The containers must be resistant, impermeable and made of a material that does not react with the product to protect it from physical, chemical and microorganism alterations. After packaging, the yoghurt should be kept refrigerated in order to increase its shelf life, which is estimated at one month.

2.2.2.4 PHYSICOCHEMICAL AND MICROBIOLOGICAL CHARACTERISTICS OF THE YOGHURT

PERUVIAN TECHNICAL STANDARD (NTP) 202.092 indicates the following:

TABLE 4
FINAL PHYSICOCHEMICAL AND MICROBIOLOGICAL CHARACTERISTICS OF YOGHURT

APPROXIMATE CHEMICAL COMPOSITION	PERCENTAGE (%)
Grease	Min. 3
Acidity (lactic acid)	0.6 -1.5
pH	4.2 – 4.6
Milk protein	Min. 2.7
Coliform count cfu/g	n m M M c 5 10 10^2 2
Fungal count cfu/g	5 10 10^2 2
Yeast cfu/g enumeration	5 10 10^2 2

Source: NTP 202.092 (2004)

2.2.3 SENSORY ANALYSIS

Ramírez-Navas (2012) indicates that the most commonly used scale is the 9-point hedonic scale, although there are also variants of this, such as the 7, 5 and 3 point scales or the smiley face graphic scale that is generally used with children. The 9-point scale is a bipolar scale. It is the recommended test for most studies, or in standard research projects, where the objective is simply to determine if there are differences between products in consumer acceptance.

Panelists are asked to evaluate coded samples of various products, indicating how much they like each sample by marking one of the categories on the scale, ranging from "extremely like" to "extremely dislike". It should be noted that the scale can be presented graphically, numerically or textually, horizontally or vertically and is used to indicate differences in consumer taste of the products. In this scale it is allowed to assign the same category to more than one sample. Samples are presented in identical containers, coded with 3-digit random numbers. The samples are coded with random numbers. The order of presentation of the samples can be randomized for each panellist or, if possible, balanced. In a balanced order of presentation, each sample is served in each of the possible positions it can occupy (first, second, third, etc.) an equal number of times, with examples of balanced layouts for 3, 4, 5 and 12 samples. An example of a hedonic test card is given in the attached figure.

FIGURE 2
9-POINT HEDONIC TEST FORM USED TO EVALUATE SENSORY ATTRIBUTES OF YOGURT.

Name :

Date: ______________________

INSTRUCTIONS

Ten samples of yogurt are presented in front of you. Please observe and taste each one, going from left to right. Indicate the degree to which you like or dislike each attribute of each sample, according to the score/category, by writing the corresponding number on the sample code line.

Score	Category	Score	Category
1	I am extremely unhappy	6	I like it slightly
2	I really dislike it	7	I like it moderately
3	I moderately dislike	8	I like it a lot
4	I slightly dislike	9	I extremely like
5	I neither like it nor dislike it		

CODE	Rating for each attribute			
	SMELL	COLOR	TASTE	TEXTURE

Source: Ligarda et al. (2012)

For data analysis, the numerical scores for each sample are tabulated and analyzed using analysis of variance (ANOVA) with Tukey's test ($\alpha = 0.05$) to determine if there are significant differences in the mean scores assigned to the samples. In the analysis of variance (ANOVA), the total variance is divided into variance assigned to different specific sources. The variance of the between-sample means is compared to the within-sample variance (also called random experimental error). If the samples are not different, the variance of the between-sample means will be similar to the experimental error. The variance corresponding to panelists or other block-grouping effects can also be compared to the random experimental error.

2.2.4 SHELF LIFE

Man and Jones (1994) indicate that essentially the shelf life of a food is defined as the time in which it will retain its physicochemical, organoleptic and nutritional properties. Shelf life encompasses several facets of nutritional value including safety, food value and sensory characteristics. When this nutritional value is affected, it significantly influences consumer purchasing decisions.
It is the time a food has before it is declared unfit for human consumption.

2.2.4.1 METHODOLOGIES FOR DETERMINING SHELF LIFE IN FOODSTUFFS

Man and Jones (1994) present methodologies for determining the shelf life of foods: prediction and assessment of shelf life:

SERVICE LIFE PREDICTION AND EVALUATION

Mathematical models and software programs to define microbiological growth and some deterioration reactions.

Real-time testing.

Accelerated testing.

SERVICE LIFE PREDICTION BY ACCELERATED METHODS

-Indispensable to know the product and its deterioration reactions.

Definition of the mechanism of the main reaction of deterioration and K-value.

-Experimentation and good correlation of the data with the Arrhenius Equation.

-Establish lifetime graphs

Correlation with sensory panels.

DIRECT METHOD

It is one of the most widely used. It involves storing the product under preselected conditions. For a period longer than the expected shelf life. Monitor periodically at regular time intervals. Observations to define the onset of deterioration.

Recommended steps:

Step 1

Identify for the specific food what may be the main possible cause of spoilage.

To know the composition of the raw material, processing aids, Aw, oxygen availability and chemical additives.

Know the possible damages related to processing, packaging and storage.

Step 2

Create a plan to establish the useful life of the product.

Time in which the study is carried out, tests and sampling dates.

-Number of samples and number of replicates

Critical environmental conditions (humidity, temperature)

Step 3

Storage of samples at the same process conditions from manufacture to consumer.

For food companies, the ability of a product to retain its full quality throughout the processing line, distribution, marketing and finally to the consumer, is the result of intensive studies to predict its shelf life.

Creating a product with a reliable shelf life requires several processes and controls by the food manufacturer.

2.2.4.2 SHELF LIFE OR SHELF LIFE ESTIMATION

Luquet (2013) indicates that when a product to be marketed is similar to others on the market, it is common practice to observe the expiration date or best before date that they present and consider a similar time for the products that we process in our companies; it is also common to resort to the bibliographic consultation of such information, however, factors that may influence a shorter or longer shelf life such as are not being considered: the quality of the raw materials, the Good Hygiene and Manufacturing Practices that were applied, the type and capacity of the processing equipment, the sanitary conditions of the facilities, the type, quality and selection of the packaging material and also the resources to maintain an adequate storage, transport and distribution of the product. Therefore, it is considered more appropriate and accurate to conduct shelf life studies designed specifically for the products. These studies can be verified in relatively short periods of time by applying the basis of accelerated shelf life studies.

There are several methodologies to perform an accelerated shelf life, however, the most common procedures, by the relative ease of its realization, are those related to the use of a factor called Q10 which is obtained considering the acceleration of biochemical or chemical reactions that a food product can present according to its composition and nature when it is stored at a temperature 10 ° C above its normal storage temperature. The value of this factor gives the guideline to calculate how many more times the shelf life of the product should be considered according to the time it maintains its stability under the accelerated conditions of the test. There are publications of studies conducted with various food products, fresh and processed, which indicate the value of Q10 estimated experimentally and these values are taken as a basis for studies of accelerated shelf life.

Once an adequate Q10 value has been selected for the product under study, the deterioration factors that it may present and therefore the selection of the analytical tests to be carried out, the protocol of the accelerated shelf life study is complemented by establishing the frequency of the analyses to be carried out. This frequency will depend on the evaluation of the speed of the changes that the product presents according to its composition, type and nature.

2.2.4.3 KEY FACTORS INFLUENCING THE SHELF LIFE OF A FOOD

Man and Jones (1994) indicate that it consists of:

- Formulation: Involves the selection of the most appropriate raw materials and functional ingredients that will ensure the integrity of the food for the required shelf life.

With respect to shelf life, key factors include moisture content, water activity (Aw), pH and addition of preservatives, antimicrobials and antioxidants.

Water activity refers to the amount of "free" water in a system available to support biological and chemical reactions; the lower the Aw, the less viable the microorganisms that contribute to product spoilage.

Preservatives belong to a class of food additives that extend shelf life by inhibiting microbial growth or minimizing the destructive effects of oxygen, metals and other factors that can lead to rancidity.

- Processing: It depends on the raw materials and ingredients to reduce unfavorable conditions or undesirable redeteriorative, promoting desirable physical and chemical changes, thus giving the food product the final form and characteristics.
- Packaging and storage conditions: The most important parameters are: relative humidity (% RH), pressure, mechanical stress, light and temperature. These parameters are dependent on both packaging and storage conditions. It is important to understand these variables in order to obtain a consistently high quality and safe food product.

2.2.4.4 ACCELERATED SHELF LIFE TESTING (ASLT)

Luquet (2013) reports that shelf-life predictions can be approached in two general ways. The most common method is to select a simple abusive condition, expose the food to that condition for various storage times, assess the quality usually by sensory methods, and then extrapolate the results to normal storage conditions. The alternative approach is to use a more elaborate design based on the principles of chemical kinetics and determine the actual temperature dependence of various quality attributes.

Accelerated methods of durability estimation are useful to reduce the time spent on estimation tests when non-perishable products are being studied.

The ASLT, try to predict the "shelf life" of a food product at a different temperature, usually higher, which allows results to be obtained in a shorter time, but with a margin of uncertainty. For this prediction to be close to the true shelf life of the food, it is necessary to know the nature of the food and its behaviour under different environmental conditions.

For the determination of shelf life of food products, by means of accelerated ASLT tests, three different storage temperatures are required, the most recommended are: T^{o}_{amb}-10°C, $T^{o}_{amb.}$ and $T^{o}_{amb+10^{o}C}$ (where T°amb is the ambient temperature), with which a graph of shelf life vs. time is obtained.

One of the most widely used models in determining the shelf life of a product is the Arrhenius Model. The Arrhenius relationship, developed theoretically for reversible molecular chemical reactions, has been experimentally applied to a number of complex chemical reactions and physical phenomena.

Food quality loss reactions have been shown to follow an Arrhenius behavior with temperature, given by:

$$K = Ko \exp\left(\frac{-E_a}{RT}\right)$$

Where:

K is the rate constant of the reaction,

Ko is the constant of the Arrhenius equation and

This is the activation energy needed to overcome product degradation.

In practical terms this means that if K values are obtained at different temperatures, and I plot LnK vs. 1/T, a straight line with slope -Ea/R is obtained. (R= 1.987cal/mol, universal gas constant).

Usually, the reaction rate is determined at three or more temperatures and K is plotted against 1/T on semilogarithmic paper or a linear regression fit of the equation is employed.

Moisture content and water activity can influence the kinetic parameters (Ea,Ko), reactant concentrations and in some cases the apparent reaction order.

Mathematical models that incorporate the effect of water activity as an additional parameter can be used for shelf life predictions of moisture-sensitive foods.

Accelerated shelf life can also be used to predict shelf life at normal conditions, based on data collected at high temperatures and high relative humidity conditions; or by means of an equation involving the activation energy, such as the Arrhenius relation

CHAPTER III

METHODOLOGY

3.1 SCOPE OR PLACE OF STUDY

The present research work was carried out in the laboratory of Nutritional Evaluation of the Faculty of Agricultural Sciences, Professional School of Agroindustrial Engineering of the National University of the Altiplano in the city of Puno, which is located at an altitude of 3827 meters above sea level, the province of Puno occupying an area of 6,492.60 km^2, within the so-called Altiplanic ecosystem between the Western and Eastern branches of the Andes Mountains, where an area of influence of Lake Titicaca is distinguished, made up of 60% pampas, plains or meadows and 40% by slopes and ravines.

The area around Lake Titicaca has favourable conditions for agriculture on gently sloping land; the grassland area has large extensions of pastures that encourage livestock activity.

The climate of the department is characterized by being cold and dry, due to its geographical location and altitude, also benefits from the thermoregulatory effect of Lake Titicaca.

3.2 POPULATION AND SAMPLE

3.2.1 POPULATION

The population for the determination of physicochemical characteristics and shelf life was constituted by the three formulations of yogurt with different levels of fiber isolated from quinoa, kiwicha and cañihua bran respectively; as well as the respective control; the same that did not have isolated bran.

While for the sensory analysis the population and sample was considered as follows:

The population consisted of 405 mothers from the food supplementation programs of the provincial municipality of Puno.

3.2.2 SAMPLE

Considering the non-probabilistic sampling using as procedure the intentional sampling or by convenience and according to the following formula the sample is:

$$n = \frac{N \times Z_a^{2} \times p \times q}{d^{2} \times (N-1) + Z_a^{2} \times p \times q}$$

$$\mathbf{n = \frac{405 \times 1.96^{2} \times 0.05 \times 0.95}{0.03^{2} \times (810-1) + 1.96^{2} \times 0.05 \times 0.95}}$$

n = 48

The sample used was 48 mothers who were beneficiaries of the PCA. From the provincial municipality of Puno.

Where:

N = population size (405 mothers in Food Supplementation Programmes)

Z = confidence level (1.96)

p = probability of success, or expected proportion (0.05)

q = probability of failure (0.95)

D = precision (Maximum permissible error in terms of proportion) (0.03)

3.3 METHODS

3.3.1 TO IDENTIFY THE PHYSICO-CHEMICAL AND MICROBIOLOGICAL CHARACTERISTICS OF THE MOST ACCEPTABLE YOGHURT WITH DIFFERENT LEVELS OF FIBRE ISOLATED FROM QUINOA, KIWICHA AND CAÑIHUA BRAN CONSIDERING THE ACCEPTABLE PARAMETERS

3.3.1.1 METHOD FOR OBTAINING THE GRAINS

Grains of the following crops were used: quinoa (Kancolla); kiwicha (Óscar Blanco), and cañihua (Cupi). The samples of quinoa and cañihua were from INIA of Puno, while the samples of kiwicha were collected from INIA of Cusco, having a representative sample of 10 kg of each of the grains indicated. Glass materials were used such as: burette, Erlenmeyer, fioles, Petri dishes, beakers, as well as other laboratory equipment and reagents recommended for the analysis of the respective methods.

3.3.1.2 OBTAINING BRAN

In order to eliminate foreign materials, the grains were cleaned (separation of foreign particles: stalks, brushwood by sieving); then the quinoa was de-stoned and dried. The grains were then ground. The bran fractions obtained were then sieved (sieve number 20) to improve the extraction yield of the bran. Bran is considered to be all that which has sizes greater than 250 µm. Finally, the samples were packed in high density polyethylene bags and conveniently stored until analysis.

3.3.1.3 FIBRE EXTRACTION (AOAC 2000)

The bran, after being digested with sulfuric acid and sodium hydroxide solutions, the residue was calcined. The difference in weight after calcination indicates the amount of fibre present.

REAGENTS

Sulfuric acid solution 0.255N.

Sodium hydroxide solution 0.313N, sodium carbonate free.

Defoamer (e.g. octyl alcohol or silicone).

Ethyl alcohol 95% (V/V).

Petroleum ether.

1% (V/V) hydrochloric acid solution.

MATERIALS AND EQUIPMENT.

Flat bottom ball flask, 600 ml, ground-glass neck.

Condensing unit for flask.

One litre Kitazato flask.

Buchner funnel.

Filtration crucible.

Rubber cones.

Whatman No. 541 filter paper.

Pizeta of 500 ml.

Desiccator.

Laboratory oven.

Muffle.

PROCEDURE 2 to 3 grams of the defatted and dried sample were weighed to the nearest milligram.

It was placed in the flask and added 200ml of the boiling sulfuric acid solution. The condenser was placed and it was taken to boiling in one minute; if necessary, antifoam was added. It was left to boil exactly for 30 min, keeping the volume constant with distilled water and periodically moving the flask to remove the particles adhered to the walls.

The Buchner funnel was installed with the filter paper and preheated with boiling water. Simultaneously and at the end of the boiling time, the flask was removed, allowed to stand for one minute and carefully filtered using suction; the filtration was done in less than 10 min. The filter paper was washed with boiling water.

The residue was transferred to the flask with the aid of a pipette containing 200ml of boiling NaOH solution and boiled for 30 min.

The filtration crucible was preheated with boiling water and carefully filtered after the hydrolysate was allowed to stand for 1 min.

The residue was washed with boiling water, with the HCl solution and again with boiling water, to finish with three washes with petroleum ether. The crucible was placed in the oven at 105°C for 12 hours and cooled in a desiccator.

The crucibles were quickly weighed with the residue (not handled) and placed in the flask at 550°C for 3 hours, allowed to cool in a desiccator and weighed again.

Calculations

A = Weight of the crucible with the dry residue (g)

B = Weight of crucible with ash (g)

C = Sample weight (g)

Crude fibre content (%)= 100((A - B)/C)

3.3.1.4 DETERMINATION OF THE SENSORY ANALYSIS (ANDALZUA METHOD 1994)

For the sensory evaluation, the consumer panel test was applied; comparison technique, proceeding as follows:

Five groups of 10 panelists each were formed; placing them at a prudent distance from each other.

Ten evaluation cards and a pencil were provided to each panelist.

Each one was given a glass with an approximate amount of 50 ml of sample A1 (3% fiber isolated from quinoa bran), instructing them to evaluate and record in the evaluation booklet. Afterwards, they were given a glass of water to neutralize the taste of the yoghurt.

Next, they were given a beaker with sample A2 (4% fibre isolated from quinoa bran), proceeding as with sample A1. Proceed in the same way with sample A3. The same procedure is performed with the yogurt samples with isolated cañihua and kiwicha fibre respectively.

For the interpretation of the results, a response sheet was used to indicate the liking or disliking of the product by means of different parameters.

3.3.1.5 pH DETERMINATION (AOAC 2002)

The equipment was calibrated with phosphate buffers of pH 4.0 and 7.0.

A volume of approximately 25.0 mL of yogurt was placed in a beaker. The electrode of the potentiometer was introduced into the sample and the pH measurement was made.

3.3.1.6 ACIDITY DETERMINATION (AOAC 2002)

20.0 mL of yogurt was measured with a volumetric pipette in an Erlenmeyer flask.

40.0 mL of CO_2 free water was added.

Three drops of phenolphthalein 1% W/V were added.

It was titrated using 0.1N sodium hydroxide until a persistent pink color was obtained.

The result was reported as percentage acidity in terms of lactic acid.

Taking into account the previous relationship, the calculations were made using the following equation:

09g = 0.009 g lactic acid = 1.0 m l NaOH 0.1 M

3.3.1.7 DETERMINATION OF FAT (AOAC 2002)

To prepare the sample homogeneously, the sample was emptied out in a beaker and the product was homogenized by shaking.

It was brought to a temperature close to 20°C.

10 mL of sulfuric acid was added to the butyrometer.

With a pipette, 11 mL of solution obtained by placing the tip of the pipette in contact with the base of the neck of the butyrometer was added carefully and very slowly, so that the neck of the butyrometer does not get wet and so that the liquids do not mix prematurely.

1 mL of amyl alcohol was poured over the surface of the mixture. The butyrometer was then closed and shaken until the casein was completely dissolved.

It was hot centrifuged for 5 min at 65 °C at 1100 rpm and the fat level was read directly on the butyrometer scale.

After direct reading on the butyrometer scale, the amount of fat in the yoghurt used in the analysis was calculated.

3.3.1.8 DETERMINATION OF MICROBIOLOGICAL ANALYSIS (MORENO 2000)

SAMPLE PREPARATION

Samples were processed within four hours after processing.

The sample was homogenized.

Dilutions of different concentrations were prepared:

10.0 mL of whey was measured and diluted in 90.0 mL of pepton water (1:10 dilution).

For the 1:100 concentration, 10.0 mL of the 1:10 dilution was measured and transferred to 90.0 mL of peptone water.

- For the 1:1000 concentration, 10.0 mL of the 1:100 dilution was measured and transferred to 90.0 mL of peptone water.

TOTAL COUNT OF MESOPHILIC AEROBIC MICROORGANISMS

1.0 mL of the 1:10 dilution was pipetted and added to a sterile Petri dish (in duplicate).

Then 15.0 mL of Plate Count agar medium was added and homogenized in a circular motion.

1.0 mL of the 1:100 dilution was added to a sterile Petri dish (in duplicate) and 15.0 mL of Plate Count agar medium was added, then homogenized in a circular motion.

1.0 mL of the 1:1,000 dilution was added to a sterile Petri dish (in duplicate) and 15.0 mL of Plate Count agar medium was added, then homogenized in a figure-eight motion.

The plates were allowed to solidify.

Plates were inverted and incubated at 30-35°C for 48 hours.

A colony counter was used to count the number of colonies.

NOTE: Each of the prepared dilutions was carried in duplicate plus a blank.

TOTAL COLIFORM COUNT BY THE MULTI-TUBE METHOD

PRESUMPTIVE EVIDENCE

Ten tubes each containing 10.0 mL of lauryl sulfate broth culture medium were collected.

2. 10.0 mL of undiluted raw yogurt/heat-treated yogurt/beverage sample was transferred to each tube.

3. The tubes were incubated at 35 °C for 48 hours.

CONFIRMATORY TEST:

1. From each tube showing gas formation, one roast was taken and inoculated into an equal number of tubes with confirmation medium (bright green bile broth).
2. It was incubated at 35 °C for 48 hours.
3. The MPN table for ten test tubes was used to interpret the results.

IDENTIFICATION OF *Escherichia coli* BY THE MULTIPLE TUBE METHOD

1. From each tube showing gas formation, a sample was taken and inoculated into an equal number of tubes with confirmation medium (EC broth).

2. Incubated for 48 hours at 45.5°C in a water bath.

3. From the tubes that had gas formation, a sample was taken and streaked on eosin methylene blue agar.

4. Incubated in oven for 18-24 hours at 35°C.

5. We proceeded to observe:

Color of the colonies, if they presented metallic green coloration there is presence of *Escherichia coli.*

Colony colour characteristics.

Greenish colonies with metallic sheen and bluish-black centre

MOLD AND YEAST COUNT

1.0 mL of yogurt was placed in triplicate in Petri dishes, using a sterile pipette for this purpose.

The procedure was repeated according to the number of dilutions prepared, using a different sterile pipette for each dilution.

15.0 to 20.0 mL of potato dextrose agar acidified with 10% W/V tartaric acid was poured, melted and kept at 45 ± 1 °C in a water bath.

The medium was carefully mixed with eight movements on a smooth surface.

The mixture was allowed to solidify by letting the Petri dishes rest on a cold horizontal surface.

A control box was prepared with 15.0 mL of medium to verify sterility. The boxes were inverted and placed in a 20°C environment.

The colonies on each plate were counted after 3, 4 and 5 days of incubation. After 5 days, those plates containing between 10 and 150 colonies were selected.

3.3.2 TO ESTABLISH THE SHELF LIFE OF THE MOST ACCEPTABLE YOGHURT WITH FIBRE ISOLATED FROM QUINOA, KIWICHA AND CAÑIHUA BRAN

3.3.2.1 DIRECT METHOD: Recommended Steps

Step 1

It was identified for the specific food what could be the main possible cause of spoilage.

To know the composition of the raw material, processing aids, Aw, oxygen availability and chemical additives.

Know the possible damages related to processing, packaging and storage.

Step 2

Create a plan to establish the useful life of the product.

Time in which the study is carried out, tests and sampling dates.

-Number of samples and number of replicates

Critical environmental conditions (humidity, temperature)

Step 3

Samples were stored at the same processing conditions from processing to consumption.

3.3.2.2 ACCELERATED TESTING:

For the determination of shelf life of the most acceptable yogurt with fiber isolated from quinoa bran, kiwicha and cañihua, through accelerated shelf life tests (ASLT) were taken at three different storage temperatures, the most recommended are: T^o_{amb}-10°C, T^o $_{amb}$. and T^o $_{amb}$+10°C (where T°amb is the ambient temperature), with which a graph of shelf life vs. time was obtained. The model used in determining the shelf life of yogurt was the Arrhenius Model.

3.3.3 TO DETERMINE THE COMPOSITION OF THE MOST ACCEPTABLE YOGHURT WITH FIBRE ISOLATED FROM QUINOA, KIWICHA AND CAÑIHUA BRAN

Determinations of the composition (moisture, protein, fat, carbohydrates, fiber, total solids and ash) of the most acceptable yogurt with fiber according to the levels of fiber isolated from quinoa, kiwicha and cañihua bran were made according to the AOAC Methods.

3.4 STATISTICAL ANALYSIS

3.4.1 TO DETERMINE THE SHELF-LIFE EFFECTS OF THE MOST ACCEPTABLE YOGHURT WITH FIBRE ISOLATED FROM QUINOA, KIWICHA AND CAÑIHUA BRAN

To determine the effects of shelf life, the sensory analyses were analyzed statistically, using the Tukey and Duncan parametric test, to determine the significance of the treatments, considering the four attributes: color, odor, flavor, texture.

To identify the physicochemical and microbiological characteristics of the most acceptable yoghurt with different levels of fibre isolated from quinoa, kiwicha and cañihua bran considering the acceptable parameters, we proceeded as follows:

Initially, the sensory analysis was carried out using a DBCA integrating the four attributes, whose model is:

$$Y_{ijk} = \mu + \alpha_i + \beta_j + (\alpha\beta)_{ij} + \varepsilon_{ijk};$$

Where:

Yijk : It is the dependent variable or response.

μ: The population mean.

αi: The effect of the i-th level of the time factor in days.

βj: is the effect of the j-th level of the factor of fibre levels.

$(\alpha\beta)_{ij}$: It is the effect of the interaction between the day factor and fibre levels.

εijk :It is the experimental error.

Subsequently the data of the most acceptable yogurt with fiber isolated from quinoa bran, kiwicha and cañihua respectively were analyzed using Tukey's and Duncan's tests.

3.4.2 TO ESTABLISH THE SHELF LIFE OF THE MOST ACCEPTABLE YOGHURT WITH FIBRE ISOLATED FROM QUINOA, KIWICHA AND CAÑIHUA BRAN

To establish the shelf life of the most acceptable yogurt with fiber according to the levels of fiber isolated from quinoa, kiwicha and cañihua bran, a DCA was used with 6 treatments (t1, t2, .. , t5, t6) and 2 replicates, the model is as follows:

$$Yij = \mu + ti + eij$$

Where:

Yij = Physicochemical characteristics observed at the ith time and in the ith repetition.

μ = Overall mean of the response variable.

ti = effect of the ith time on the response variable

eij = Experimental error

3.4.3 TO DETERMINE THE COMPOSITION OF YOGHURT WITH FIBRE ISOLATED FROM QUINOA, KIWICHA AND CAÑIHUA BRAN

To determine the most acceptable yogurt composition with different levels of fiber isolated from quinoa, kiwicha and cañihua bran during shelf life, descriptive statistics were used as measures of central tendency: average and measures of dispersion (standard deviation), whose formulas are: Average $\mu = \frac{\sum X}{N}$

Where:

μ = Average

$\sum$ = Summation

N = Number of elements

X = Values or data

Standard deviation $S = \sqrt{S^2}$

Where:

S = Standard deviation

S2 = Variance

CHAPTER IV

RESULTS AND DISCUSSIONS

To identify the physicochemical and microbiological characteristics of the most acceptable yoghurt with different levels of fibre isolated from quinoa, kiwicha and cañihua bran, a sensory analysis was carried out, the results of which are shown below.

TABLE 5
AVERAGES AND STANDARD DEVIATIONS OF THE SENSORY ANALYSIS OF YOGURT WITH ISOLATED QUINOA BRAN FIBER. PUNO JANUARY - MARCH 2015

Addition (%)	Average	± Standard Deviation
3	6.005	± 1.44
4	6.267	± 1.395
5	6.792	± 1.167

Source: Data obtained from the research work

TABLE 6
DUNCAN COMPARISON TEST FOR THE LEVELS OF FIBER ISOLATED FROM QUINOA BRAN. PUNO JANUARY - MARCH 2015

Order of Merit	Additions (%)	Average	N. S
1	5	6.792	a
2	4	6.266	b
3	3	6.005	b

Source: Data obtained from the research work

TABLE 7
AVERAGES AND STANDARD DEVIATIONS OF SENSORY ANALYSIS OF YOGHURT WITH FIBRE ISOLATED FROM CAÑIHUA BRAN. PUNO JANUARY-MARCH 2015

Addition (%)	Average	+Standard deviation
5.	6.7457	± 1.435
4	6.708	± 1.398
3	6.505	± 1.167

Source: Data obtained from the research work

TABLE 8
DUNCAN'S COMPARISON TEST FOR YOGHURT WITH LEVELS OF FIBRE ISOLATED FROM CAÑIHUA BRAN. PUNO JANUARY - MARCH 2015

Order of Merit	Additions (%)	Average	N.S
1	5	6.745	a
2	4	6.708	a
3	3	6.505	b

Source: Data obtained from the research work

TABLE 9
AVERAGES AND STANDARD DEVIATIONS OF THE SENSORY ANALYSIS OF YOGHURT WITH FIBRE ISOLATED FROM KIWICHA BRAN, PUNO JANUARY - MARCH 2015

Addition (%)	Average	Standard deviation
5	6.745	± 1.435
4	6.708	± 1.395
3	6.505	± 1.167

Source: Data obtained from the research work

TABLE 10
DUNCAN COMPARISON TEST FOR KIWICHA ISOLATED FIBER LEVELS. PUNO JANUARY - MARCH 2015

Order of Merit	Additions (%)	Average	N S
1	5	6.932	a
2	4	6.823	b
3	3	6.489	c

Source: Data obtained from the research work

Tables 1, 2, 3, 4 and 5 show the results of the sensory evaluation carried out by the panelists on the attributes of smell, color, flavor and texture of yogurt with the addition of quinoa, cañihua and kiwicha fiber with 3 percentages: 3, 4 and 5 grams per 100 ml, the yogurt that had greater acceptance is the one that had 5g/100 of fiber isolated from quinoa bran, kiwicha and cañihua respectively.

This is explained by Diaz-Jimenez (2004) who in ANOVA found significant differences in the judges' perception of wheat bran content and storage time. Among the treatments to which wheat bran was incorporated, the best rated was the one with the lowest addition (1%).

Similar results were found for cow's milk yogurt with added fiber. Likewise, it is observed that the treatment with less fiber addition is the most similar to the treatment without bran addition, which suggests the need to mask the flavor provided by the fiber by means of a flavoring agent. In the case of the yogurt obtained in this study, there was no need to mask the product with any additive.

4.1 IDENTIFICATION OF THE PHYSICOCHEMICAL AND MICROBIOLOGICAL CHARACTERISTICS OF THE MOST ACCEPTABLE YOGHURT WITH DIFFERENT LEVELS OF FIBRE ISOLATED FROM QUINOA, KIWICHA AND CAÑIHUA BRAN.

The physicochemical and microbiological characteristics of the most acceptable yoghurt with fibre isolated from quinoa, kiwicha and cañihua bran are shown below.

TABLE 11
PHYSICOCHEMICAL CHARACTERISTICS OF THE MOST ACCEPTABLE YOGHURT WITH ISOLATED FIBRE FROM QUINOA, KIWICHA AND CAÑIHUA BRAN. PUNO JANUARY - MARCH 2015

Yogurt with bran fiber isolate	Physicochemical characteristics				
	pH	Acidity	Protein (g)	Fat (g)	Fiber (B.S)
QUINUA	5.3	0.7	4.1	3	0.3
KIWICHA	5.1	0.7	3.3	2.4	0.4
CAÑIHUA	5.2	0.7	3.8	2.5	0.3
CONTROL	5.1	0.8	3.4	2.8	0.0

Source: Data obtained from the research work

This table shows that in terms of pH and acidity the yogurt with isolated quinoa fiber has (5.31 and 0. 735 respectively), protein (4.05%) and fat (3%); the yogurt with isolated kiwicha fiber has less protein (3.26%), fat (2.4%) and higher fiber expressed on dry basis (0.4%), yogurt with isolated fiber cañihua higher acidity (0.705) and lower fiber on dry basis (0.3) along with quinoa with the same proportion. With respect to yoghurt without the addition of isolated fibre, it has the highest proportion of acidity. This is explained by Simanca et al. (2013) who found that the best evaluated treatments with a higher percentage of fiber had higher pH and lower acidity values.

A study of four quinoa varieties showed that dietary fibre in raw quinoa ranged from 13.6 g to 16.0 g per 100 g dry weight Repo-Carrasco et al. (2003). Most of the dietary fibre was insoluble, ranging from 12.0 g to 14.4 g compared to 1.4 g to 1.6 g of soluble fibre per 100 g dry weight. Similar to the total protein value of quinoa, the value of dietary fibre is generally higher than most grains and lower than legumes. Dietary fibre is the part of plant foods that cannot be digested and is important for facilitating digestion and preventing faecal stagnation in the intestine.

TABLE 12
MICROBIOLOGICAL CHARACTERISTICS OF THE MOST ACCEPTABLE YOGHURT WITH FIBRE ISOLATED FROM QUINOA, KIWICHA AND CAÑIHUA BRAN. PUNO JANUARY - MARCH 2015

Yogurt with bran fiber isolate	Microbiological characteristics (cfu/g)			
	Viable mesophilic aerobes	Coliforms	Yeast	Molds
Quinoa	$1,60 \times 10^6$	Neg.	$3,4 \times 10^4$	Neg.
Kiwicha	$4,2 \times 10^5$	Neg.	$1,1 \times 10^5$	Neg.

Cañihua	6,2 X 10^5	Neg.	2,7 X 10^4	Neg.
Control	2,10 X 10^5	Neg.	4,8 X 10^4	Neg.

Source: Data obtained from the research work

In the present study, only viable mesophiles were found within the established parameters; coliforms and molds were absent, which guarantees the application of the GMP. The data found are supported by the studies of Orbera (2004) who reported that the yogurt produced complies with the microbiological requirements which is attributed to the adequate pasteurization of the milk, guaranteeing a good quality product. On the other hand, the storage conditions contributed to the microbiological stability of the yoghurt. The formulated yogurt did not present total coliforms which indicates the adequate hygienic quality with which it was elaborated. In general, fermented dairy products are free of pathogens and the prebiotic strains in the yogurt obtained seem to accentuate the inhibitory effect of this product on some pathogenic bacteria.

Yeasts possess certain particular characteristics that allow them to grow and contaminate in dairy foods, including lactose fermentation/assimilation, production of extracellular proteolytic enzymes, e.g. lipases, assimilation of lactic and citric acid, growth at low temperatures.

In milk, yeast contamination can occur after pasteurization and is a secondary contamination, involving *Cry. flavus*, *Cry. diffluens*, *Deb. hansenii* and *Kluyveromyces marxianus* species. Refrigerated raw milk tolerates the growth of *Cry. curvatus*, *G. candidum*, *Deb. hansenii*, among others.

One of the dairy products mostly altered by the action of yeasts is yoghurt, due to the addition of fruits and flavourings derived from fruits.

4.2 SHELF LIFE OF THE MOST ACCEPTABLE YOGHURT WITH FIBRE ISOLATED FROM QUINOA, KIWICHA AND CAÑIHUA BRAN.

The following figures show the physicochemical characteristics (pH, acidity, protein, fat and fibre) that will allow us to establish the shelf life of the most acceptable yoghurt with isolated fibre from quinoa, kiwicha and cañihua respectively.

FIGURE 3
pH OF THE MOST ACCEPTABLE YOGHURT WITH FIBER ISOLATED FROM QUINOA, KIWICHA AND CAÑIHUA SAVAGE, FROM 0 TO 35 DAYS. PUNO JANUARY - MARCH 2015.

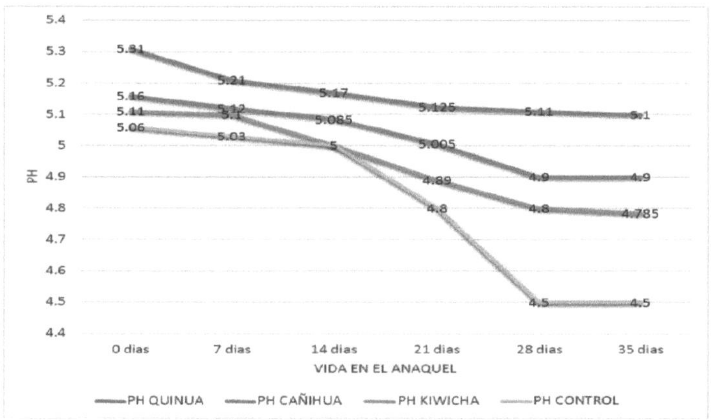

Source: Data obtained from the research work

This figure shows the pH of yogurt under study, as a function of time (0, 7, 14, 21, 28 and 35 days), in which a decrease is observed. When performing the respective statistical analysis, there is a significant difference, as shown in Annex 4.

Regarding the shelf life for this parameter (pH) by means of the direct method, it is shown that the yogurt formulations under study decrease progressively. In this regard Diaz-Jimenez (2004), indicates that the fiber would exert an influence on the pH, both of freshly made yogurt and during storage, was above 4; although in each system there are differences, this variation responds mainly to two aspects: to the influence exerted by the fiber on the pH of yogurt, to the production of lactic acid by lactic acid bacteria during storage of each system, and to the breaking of the gel in the case of whipped yogurt. The lowest pH values (4.4) were recorded at the end of the study period (21 days) and the decreasing evolution of pH was similar for both types of yogurt. It is considered that both fiber and fat provide acid elements; and on the other hand, lactic acid is produced by lactic acid bacteria during storage.

On the other hand, Rojas et al. (2007) indicate that the phenomenon of stabilization of the pH of the samples at the end of storage is probably due to the inhibition of the enzymatic activity of the cultures and the decrease in the bacterial load, although it is probably also due to the depletion of the available lactose reserves. However, it may be thought from the storage conditions (especially the low temperature) that inhibition of enzyme activity and decline in microorganisms is more likely to occur first, before the lactose is completely used up.

FIGURE 4
ACIDITY OF THE MOST ACCEPTABLE YOGURT WITH FIBER ISOLATED FROM QUINOA, KIWICHA AND CAÑIHUA BRAN, FROM 0 TO 35 DAYS. PUNO JANUARY - MARCH 2015

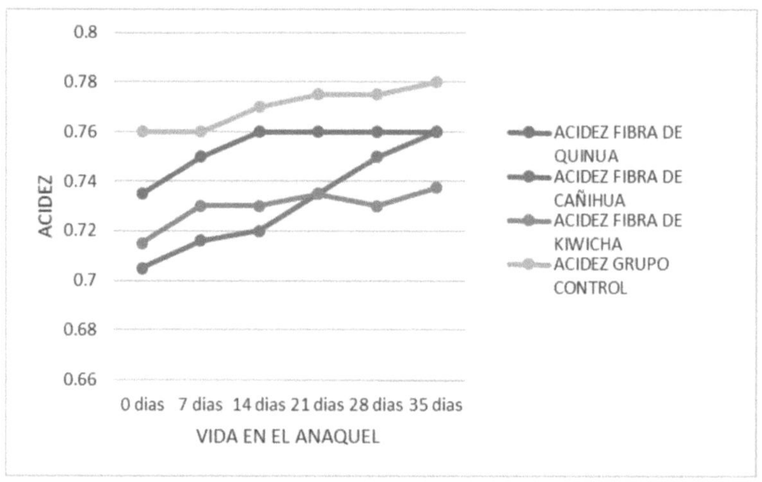

Source Data obtained from the research work

This figure shows the acidity of the yogurt under study, as a function of time (0, 7, 14, 21, 28 and 35 days), in which an increase is observed.

The variation of acidity was higher for the yogurt with fiber isolated from hemp bran, this means that from day 14 the acidity values reach 0.76 on day 35. When the respective statistical analysis is done, there is a significant difference.

Regarding the shelf life for this parameter (acidity) by the direct method and the accelerated shelf life tests (ASLT) it is shown that yogurt with fiber isolated from cañihua bran increases from day 21 onwards. These data are explained by Díaz-Jiménez (2004), who indicates that the acidity of yogurt, which is intimately and inversely related to the decrease in pH, increased during storage, from 0.65% for freshly made samples to 1.1% for samples stored for three weeks. *(*1992) cited by Diaz-Jimenez (2004), who quantified an acidity of 0.7 to 1.35% in soy yogurt; and Pirkul *et al.* (1997), who report an acidity between 0.85 and 1.1 for different formulations of yogurt enriched with calcium and stored for 14 days. The changes in acidity are basically the result of biochemical transformations present in the yoghurt during its shelf life. Statistical analyses indicate that there is only significant effect of storage time and not of composition (neither fiber nor fat); which as mentioned above, can be attributed mainly to the activity of lactic acid producing lactic acid bacteria during the storage period studied. The average values of three acidity determinations for all the systems studied with their corresponding deviations, throughout storage.

In the studies of Tola (2006) the parameter of acidity (Q) of yogurt, is the one that has been chosen for the experimental monitoring of the kinetics of deterioration as a function of time and temperature, having determined that the adjustment of data to the kinetics of zero order and the Arrhenius equation, according to the interpretation of the value of the coefficient of determination and regression graphs, acceptable, within the limits of experimentation developed. The values found for the shelf life of yogurt in its different characteristics (with preservative, without preservative and natural), establish an inverse relationship between the storage temperature and shelf life (the higher the temperature the shorter the shelf life), on the other hand, the influence of the use of preservative is decisive to extend the shelf life of yogurt. Finally, the non-addition of preservative involves a rapid rise in acidity between the end of the preparation and the first day of control.

FIGURE 5
PROTEIN OF THE MOST ACCEPTABLE YOGURT WITH FIBER ISOLATED FROM QUINOA, KIWICHA AND CAÑIHUA BRAN, FROM 0 TO 35 DAYS. PUNO JANUARY - MARCH 2015

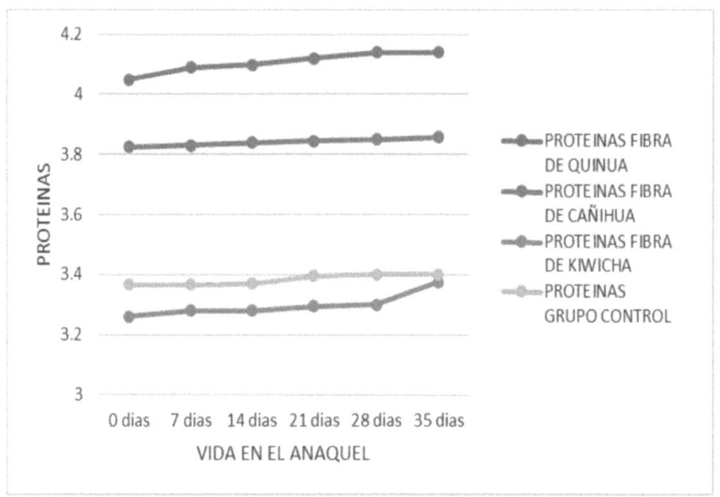

Source: Data obtained from the research work

This figure shows the protein content of the yogurt under study, as a function of time (0, 7, 14, 21, 28 and 35 days), in which an increase is observed. The ANOVA shows that there is a significant difference.

Regarding the shelf life for this parameter (protein) by the direct method, the protein content does not show considerable variability. The function of proteins, by improving the texture, also masks the acidity, and the fat provides a smoother and creamier flavor and a better aroma as Luquet (2013) argues. The quantity of protein in quinoa depends on the variety, with a range between 10.4% and 17.0% of its edible part. Although it generally has a higher amount of protein in relation to most grains, quinoa is known more for the quality of the protein as stated by Repo-Carrasco et al (2003) The protein is composed of amino acids, eight of which are considered essential for both children and adults. Unlike quinoa, most grains are low in the essential amino acid lysine, while most legumes are low in the sulphur amino acids methionine and cysteine.

FIGURE 6
FAT OF THE MOST ACCEPTABLE YOGURT WITH FIBER ISOLATED FROM QUINOA, KIWICHA AND CAÑIHUA BRAN, FROM 0 TO 35 DAYS. PUNO JANUARY - MARCH 2015

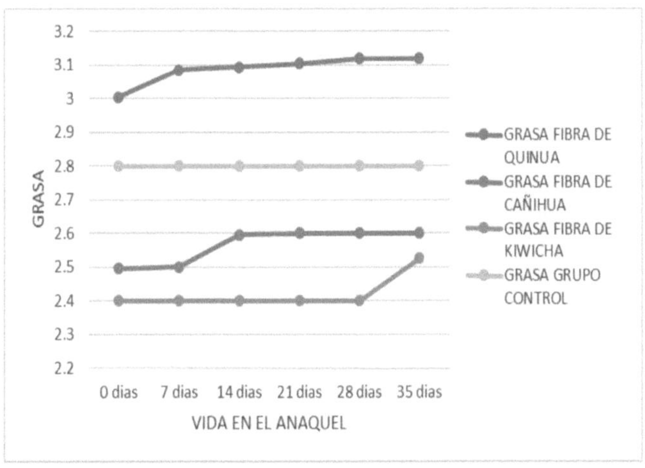

Source: Data obtained from the research work

This figure shows yogurt proteins under study, as a function of time (0, 7, 7, 14, 21, 28 and 35 days), in which an increase is observed. The ANOVA shows that there is a significant difference.

As for the shelf life for this parameter (fat) by the direct method, from day 28 to day 35 shows a considerable increase.

Fats are an important source of calories and facilitate the absorption of fat-soluble vitamins. Of the total fat content of quinoa, more than 50% comes from the essential polyunsaturated fatty acids linoleic (omega 6) and linolenic (omega 3). Linoleic and linolenic acids are considered essential fatty acids, as they cannot be produced by the body. It has been shown that the fatty acids in quinoa maintain the quality due to the high natural value of vitamin E, which acts as a natural antioxidant as explained by Repo-Carrasco et al. (2003)

FIGURE 7
FIBER (DRY BASIS) OF THE MOST ACCEPTABLE YOGURT WITH ISOLATED FIBER FROM QUINOA, KIWICHA AND CAÑIHUA BRAN, FROM 0 TO 35 DAYS. PUNO JANUARY - MARCH 2015

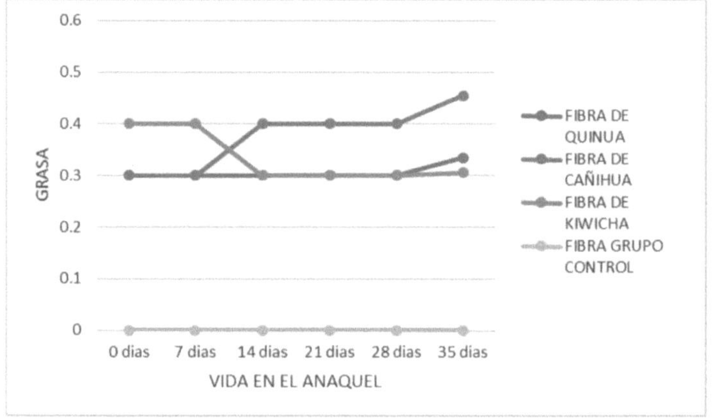

Source: Data obtained from the research work

The present figure shows fiber (dry basis) of yogurt under study, as a function of time (0, 7, 14, 21, 28 and 35 days), in which an increase is observed for yogurt with fiber isolated from hemp bran from day 7, stabilizing until day 28, and then again showing an increase at day 35. The ANOVA shows that there is a significant difference. This is supported by Peña (2005), quoted by Repo-Carrasco et al. (2003) who indicate that quinoa is a source of insoluble fibre, with a higher concentration in the cooked grains than in the raw ones, possibly because some processes induce the formation of less digestible compounds such as starch-tannin complexes and products of the Maillard reaction, which become part of the insoluble fibre.

4.3 COMPOSITION OF THE MOST ACCEPTABLE YOGHURT WITH DIFFERENT LEVELS OF FIBRE ISOLATED FROM QUINOA, KIWICHA AND CAÑIHUA BRAN.

TABLE 13
COMPOSITION OF THE MOST ACCEPTABLE YOGHURT WITH ISOLATED FIBRE FROM QUINOA, KIWICHA AND CAÑIHUA BRAN. PUNO JANUARY - MARCH 2015

Yogurt with bran fiber isolate	Composition (g/100 g)						
	Humidity (%)	Prot.	Grease	Carboh.	Fiber (BS)	Total Solids	Ash
Quinoa	76.9	4.1	3.1	14.7	0.3	23.1	0.8
Kiwicha	76.6	3.4	2.5	16.5	0.3	23.4	0.7
Cañihua	77.9	3.9	2.6	14.5	0.5	21.1	0.8
Control	80.3	3.4	2.8	12.8	0.0	19.7	0.7

Source: Data obtained from the research work

This table shows the composition of the most acceptable yogurt with isolated fiber from quinoa, kiwicha and cañihua bran; in which the yogurt with isolated fiber from cañihua bran has 77.9% moisture, protein, fat yogurt with isolated fiber from quinoa bran has 4.1 g and 3.1 g respectively; carbohydrates the yogurt with isolated fiber of kiwicha bran presents 16.5 g; fiber (dry basis) the yogurt with isolated fiber of cañihua bran presents 0.5 g; total solids the yogurt with isolated fiber of kiwicha bran presents 23.4 g and ash the yogurt with isolated fiber of quinoa bran presents 0.8 g.

A recent study of four quinoa varieties showed that dietary fibre in raw quinoa ranged from 13.6 g to 16.0 g per 100 g dry weight Repo-Carrasco et al. (2003). Most of the dietary fibre was insoluble, ranging from 12.0 g to 14.4 g compared to 1.4 g to 1.6 g of soluble fibre per 100 g dry weight. Similar to the total protein value of quinoa, the value of dietary fibre is generally higher than most grains and lower than legumes. Dietary fibre constitutes the part of plant foods that cannot be digested and is important for facilitating digestion and preventing faecal stagnation in the gut. This is sustained by Simanca et al. (2013) who found results that agree with those obtained by Atta et al. (2009) in buffalo milk yogurt with different concentrations of starter culture, and with those obtained in the physicochemical analysis, since the treatments with lower addition of fiber presented higher pH and lower values of acidity and syneresis; but are similar to those obtained in the sensory aspect in natural yogurt made with buffalo milk during storage time as sustained by Cunha et al. (2005). This fact can be attributed to the effect exerted by the acidity, which masks the flavor of wheat bran, being the best qualified with the lowest addition (1%). Similar results were found in cow's milk yogurt added with fiber, reported by Diaz et al., 2004. Likewise, it is observed that the treatment with the lowest fiber addition is the most similar to the treatment without bran addition, which suggests the need to mask the flavor provided by the fiber by means of a flavoring agent.

CONCLUSIONS

The physicochemical and microbiological characteristics show that in terms of pH, protein and fat the yoghurt with isolated fibre from quinoa and kiwicha bran was found to have 5.3 and 5.1, 4.1 and 3.3%; 3 and 2.4% respectively; while for acidity, the yoghurt with isolated fibre from quinoa and kiwicha had 0.7 and 0.7%; while for fibre, the yoghurt with isolated fibre from kiwicha bran had 0.4 and 0.3% for quinoa and cañihua.

The microbiological characteristics (cfu/g) the presence of viable mesophilic aerobic bacteria is 1.60 X 10^6, 4.2 X 10^5 and 6.2 X 10^5; yeasts 3.4 X 10^4, 1.1 X 10^5 and 4.8 X $10^{4,}$ not finding coliform bacteria or molds respectively.

As for shelf life:

- pH at day zero: 5.31, 5.11 and 5.16; and at day 35: 5.1, 4.785 and 4.9.
- Acidity on day zero: 0.735, 0.715 and 0.705 and on day 35 is 0.76, 0.7375 and 0.7375.

The composition of yogurt with isolated fiber from quinoa, kiwicha and cañihua bran is:

- Humidity: 76.9, 76.6 and 77.9%.
- Protein: 4.1, 3.4 and 3.9%.
- Fat: 3.1, 2.5 and 2.6%.
- Carbohydrates: 14.7, 16.5 and 14.5.
- Fiber: 0.3, 23.4 and 0.5%.
- Total Solids: 23.1, 23.4 and 22.1
- Ash: 0.8, 0.7 and 0.8%.

RECOMMENDATIONS

Conduct similar studies considering variability of yoghurt (plain, drinkable, etc.).

Obtain fiber from other grains and incorporate them into foods such as yogurt, nectars, among others.

BIBLIOGRAPHY

Acevedo I, García O, Contreras J. (2007) *Characteristics of strawberry whipped yogurt derived from different proportions of cow and goat milk.* San José, Costa Rica. University of Costa Rica. Volume 18 Nº pp 2221-237

Anzaldua, A (2004*) Sensory evaluation of food in theory and practice.* Edit. Acribia Zaragoza Spain

Apolinario I (2014) *Review Inulin-type fructans: A review on different aspects of biochemical and pharmaceutical technology. Carbohydrate Polymers*

AOAC (2002) *Official Methods of Analysis.* 17th edition. Ed. Association of Official Analytical Chemists, international Gaitherstourg, USA.

Bueno-Solano C, Campas-Baypoli O, Diaz-Garcia A, Izaguirre-Flores I, Verdugo-Zamorano W. (2009) *Quantification of riboflavin (vitamin B2) in dairy products by HPLC.* Chile. Rev Chil Nutr Vol. 36, No. 2, pp: 136-142.

Cunha O, Oliveira R, Hotta Y, Sobral P, (2005) *Avaliação físico-química e sensorial do iogurte natural produzido com leite de búfala contendo diferentes níveis de gordura*, Ciência e Tecnologia de Alimentos. São Paulo Brazil.

Díaz-Jiménez B. (2004*) Effect of fiber addition and fat reduction on physicochemical properties of yogurt.* Department of Chemical Engineering and Food Engineering, Universidad de las Américas, Puebla Mexico. Mexican journal of chemical engineering. Vol 3 Nº 3 pp 287-305.

Garcia P (2000) *Fibre and health. Nutrition and obesity*. Zaragoza Spain

Gibson, G (2004*) Effects on probiotics and fibre.* Madrid Spain

Hernández A. (2003) *Industrial Microbiology*. Editorial UNED.

Ilsi E. (2006) *Dietary fibre: definition, analysis, physiology and health*. Brussels Belgium

Kin (2000) *Impact of dietary fiber on the occurrence of colon cancer.* Gastroenterology. Zaragoza Spain

Ligarda C, Repo-Carrasco R, Encina C, Herrera I, (2012) *Extraction with neutral and alkaline solutions for the isolation of soluble and insoluble fiber from the bran of quinoa* (*Chenopodium quinoa* Willd,), *Kiwicha (Amaranthus caudatus L,), and Cañihua (Chenopodium pallidicaule Aellen*). Lima. Rev. Soc. quim. Peru. Vol 78 Nº 1 pp 53-64

Luquet, F. (2013). *Milk and dairy products. Cow - Sheep - Goat.* Zaragoza Spain. Editorial Acribia, S.A.

Lubbers S (2004). *Flavor release and rheology behaviour of strawberry fat free stirred yogurt during storage*. Journal Agricultural Food Chemistry. Vol 52 No. 10 pp 77 - 82

Man C and Jones A (1994) *Shelf life evaluation of foods*. Blackie Academic and Professional, London.

Mataix J. (2006) *Nutrition and human nutrition*. Ediciones Díaz de Santos Madrid. Spain

Miranda O, Fonseca P, Ponce I, Cedeño C, Sam I, Marti I. (2014) *Elaboration of a fermented beverage from whey incorporating Lactobacillus acidophilus and Streptococcus thermophilus*. Cuban Journal of Food and Nutrition Vol 24 No. 1 pp 7-16.

Moreno B (2000) *International Commission on Microbiological Specifications for Foods (ICMSF)*. Edit. Acribia Zaragoza Spain.

Olivares S, Andrade, M. and Zacarías, I. (1994) *Necesidades nutricionales y calidad de la dieta; Manual de auto instrucción*. Santiago, Universidad de Chile. Institute of Nutrition and Food Technology.

Orbera T. (2004) *Harmful action of yeasts on food*. Cuban journal of public health.

Ramírez-Navas (2012). *Sensory Analysis: Consumer-oriented tests.* Universidad del Valle Cali Colombia

Repo-Carrasco, R., Espinoza, C. and Jacobsen, S.E. (2003) *Nutritional value and use of the Andean crops quinoa (Chenopodium quinoa Willd.) and kañiwa (Chenopodium pallidicaule).* Food Reviews International. Vol. 19, Nos. 1 and 2.

Rodríguez V, Cravero F, Alonso A, (2002*) Process of elaboration of lactose-free yogurt from goat milk.* Córdoba Argentina

Rojas-Castro, W, Chacón A, Pineda-Castro M. (2007) *Characteristics of strawberry whipped yogurt derived from different proportions of cow and goat milk.*

Romero R (2008). *Dairy products technology.* Ediciones UPC.

Rubioma (2002): *Implications of fibre in different pathologies.*

Ruiz, J, Rivera A, Ramírez M, (2009) *Elaboration of yogurt with probiotics (Bifidobacterium spp. and Lactobacillus acidophilus) and inulin.* Institute of Chemistry and Technology, Faculty of Agronomy, Central University of Venezuela. Journal of the Faculty of Agronomy.

Sastre (2008) *Fibre and prebiotics: concepts and perspectives.* Gastroenterología Madrid Spain.

Simanca, M, Andrade R. Arteaga R. (2013). *Effect of wheat bran on the physicochemical and sensory properties of buffalo milk yogurt.* Technological information. Vol 24. N° 1 pp 79-86

Tola, F. (2006), *Determination of the shelf life of yogurt.* Technical University of Oruro Bolivia

Valenzuela and Maíz. (2006). *The role of dietary fiber in enteral nutrition.* Rev. Chil. Nutr. Vol. 33, Supplement N°2, pp 342-351.

ANNEXES

ANNEX 1

PHYSICOCHEMICAL CHARACTERISTICS SHEET OF YOGHURT WITH ISOLATED BRAN FIBRE

Product	Sample Code	Date	Physicochemical characteristics			
			pH	Acidity	Grease	Fiber

ANNEX 2

MICROBIOLOGICAL CHARACTERISTICS SHEET FOR YOGHURT WITH ISOLATED BRAN FIBRE

Product	Sample Code	Date	Microbiological characteristics			
			Mesophilic aerobes	Molds	Yeast	Coliforms

APPENDIX 3
SENSORY ANALYSIS FORM

Name :

Date:

INSTRUCTIONS

Ten samples of yogurt are presented in front of you. Please observe and taste each one, going from left to right. Indicate the degree to which you like or dislike each attribute of each sample, according to the score/category, by writing the corresponding number on the sample code line.

Score	Category	Score	Category
1	I am extremely unhappy	6	I like it slightly
2	I really dislike it	7	I like it moderately
3	I moderately dislike	8	I like it a lot
4	I slightly dislike	9	I extremely like
5	I neither like it nor dislike it		

CODE	Rating for each attribute			
	SMELL	COLOR	TASTE	TEXTURE

Figure. 9-point hedonic test card used to evaluate sensory attributes of yogurt.

ANNEX 4
ANNOTATIONS OF THE SENSORY ANALYSIS

ANOVA OF THE SENSORY ANALYSIS OF YOGURT WITH FIBER ISOLATED FROM QUINOA BRAN.

F. of V.	G.L.	S.C.	C.M.	Fc	Meaning
A: Judges	49	375.917	7.67176	6.85567	**
B: Treatments	2	14.7925	7.39625	6.60946	**
Interaction A x B	1	2.97686	2.97686	2.66019	
Experimental error	98	109.665	1.11903		
Total	149	500.375			

Source: Data obtained from the research work

ANOVA OF SENSORY ANALYSIS OF YOGURT WITH FIBER ISOLATED FROM CAÑIHUA BRAN.

F. of V.	G.L.	S.C.	C.M.	Fc	Meaning
A: Judges	49	483.6567	9.87054	16.6761	**
B: Treatments	2	1.535833	0.76791	1.29738	
AxB Interaction	1	3.258043	3.25804	5.50441	**
Experimental error	98	58.00583	0.59189		
Total	149	543.1983			

Source: Data obtained from the research work

ANOVA OF THE SENSORY ANALYSIS OF YOGHURT WITH ISOLATED KIWICHA FIBRE.

F. of V.	G.L.	S.C.	C.M.	Fc	Meaning
A: Judges	49	407.7838	8.32212	11.3577	**
B: Treatments	2	4.900833	2.45042	3.34423	
Interaction A x B	1	3.351237	3.35123	4.57358	
Experimental error	98	71.80750	0.73273		
Total	149	484.4920			

Source: Data obtained from the research work

EVALUATION OF THE SENSORY PROPERTIES OF YOGURT ADDED WITH DIFFERENT PERCENTAGES OF QUINOA, CAÑIHUA AND KIWICHA FIBERS

AVERAGES AND STANDARD DEVIATIONS OF THE SENSORY ANALYSIS OF YOGURT ADDED WITH FIBER ISOLATED FROM QUINOA BRAN.

ADDITION (%)	AVERAGE	STANDARD DEVIATION
3	6.00520833	± 1.4352
4	6.26562500	± 1.3957
5	6.79166667	± 1.1673

Source: Data obtained from the research work

ANOVA is shown, using the Randomized Complete Block Design (RCBD).

ANOVA OF THE SENSORY ANALYSIS OF YOGURT ADDED WITH FIBER ISOLATED FROM QUINOA BRAN.

F. of V.	G.L.	S.C.	C.M.	Fc	Meaning
A: Judges	49	375.917	7.67176	6.85567	**
B: Treatments	2	14.7925	7.39625	6.60946	**
Interaction A x B	1	2.97686	2.97686	2.66019	
Experimental error	98	109.665	1.11903		
Total	149	500.375			

Source: Data obtained from the research work

DUNCAN COMPARISON TEST FOR JUDGES WHO TASTED YOGURT WITH FIBER ISOLATED FROM QUINOA BRAN.

ORDER OF MERIT	JUDGES	AVERAGE	SIGNIFICANCE LEVEL
1	33	8.16667	a
2	32	8.16667	a
3	48	7.91667	b
4	47	7.83333	b
5	31	7.58333	c
6	11	7.58333	c
7	35	7.41667	d
8	36	7.33333	d
9	9	7.25000	e
10	8	7.25000	e
11	26	7.25000	e
12	46	7.16667	e
13	38	7.08333	f
14	45	7.00000	f
15	40	6.91667	fg
16	3	6.91667	fg
17	37	6.83333	g
18	21	6.83333	g
19	44	6.75000	h
20	27	6.75000	h
21	42	6.66667	hi
22	7	6.58333	i
23	12	6.58333	i
24	34	6.50000	j
25	25	6.41667	j
26	41	6.33333	jk
27	39	6.25000	k
28	29	6.25000	k
29	43	6.08333	l
30	20	6.00000	lm
31	4	5.91667	m
32	28	5.91667	m
33	19	5.91667	m
34	23	5.75000	m n
35	17	5.66667	n
36	22	5.58333	n
37	18	5.58333	n
38	2	5.50000	o
39	10	5.50000	o
40	13	5.41667	op
41	1	5.33333	p
42	16	5.25000	p
43	5	5.08333	p
44	15	5.00000	p
45	24	4.83333	q
46	14	4.75000	q
47	30	4.66667	r

48	6	3.66667	r

Source: Data obtained from the research work

DUNCAN COMPARISON TEST FOR QUINOA BRAN ISOLATED FIBRE ADDITIONS

ORDER OF MERIT	ADDITIONS (%)	AVERAGE	SIGNIFICANCE LEVEL
1	3	6.79166667	a
2	4	6.26562500	b
3	5	6.00520833	b

Source: Data obtained from the research work

AVERAGES AND STANDARD DEVIATIONS OF THE SENSORY ANALYSIS OF YOGURT ADDED WITH FIBER ISOLATED FROM CAÑIHUA BRAN.

ADDITIONS	AVERAGE	STANDARD DEVIATION
203	6.74479167	± 1.4352
202	6.70833333	± 1.3957
201	6.50520833	± 1.1673

Source: Data obtained from the research work

ANOVA, USING THE RANDOMIZED COMPLETE BLOCK DESIGN (DBCA).

ANOVA OF THE SENSORY ANALYSIS OF YOGURT ADDED WITH FIBER ISOLATED FROM CAÑIHUA BRAN.

F. of V.	G.L.	S.C.	C.M.	Fc	Meaning
A: Judges	49	483.6567	9.87054	16.6761	**
B: Treatments	2	1.535833	0.76791	1.29738	
AxB Interaction	1	3.258043	3.25804	5.50441	**
Experimental error	98	58.00583	0.59189		
Total	149	543.1983			

Source: Data obtained from the research work

DUNCAN COMPARISON TEST FOR JUDGES WHO TASTED YOGURT WITH FIBER ISOLATED FROM CAÑIHUA BRAN.

ORDER OF MERIT	JUDGES	AVERAGE	SIGNIFICANCE LEVEL
1	8	8.66667	a
2	6	8.33333	b
3	18	8.33333	b
4	16	8.33333	b
5	11	8.33333	b
6	7	8.16667	c
7	28	8.08333	d
8	1	8.08333	d
9	36	8.00000	e
10	25	8.00000	e
11	2	7.91667	e
12	5	7.83333	ef
13	23	7.58333	f
14	14	7.58333	f
15	13	7.58333	f
16	4	7.50000	g
17	20	7.33333	g
18	9	7.25000	h
19	10	7.25000	h
20	43	7.16667	i
21	15	7.08333	i
22	29	6.91667	j
23	24	6.91667	j
24	22	6.66667	k
25	3	6.58333	kl
26	27	6.50000	l
27	21	6.50000	l
28	17	6.50000	l
29	32	6.41667	m
30	19	6.33333	m
31	35	6.25000	n
32	34	6.25000	n
33	12	6.25000	n
34	26	6.16667	n o
35	38	6.00000	o
36	44	5.91667	o p
37	33	5.83333	p
38	31	5.75000	q
39	47	5.33333	r
40	41	5.16667	r
41	39	5.08333	r
42	30	5.08333	r
43	46	4.83333	r s
44	45	4.58333	s
45	40	4.58333	s
46	37	4.41667	s t
47	42	4.08333	t
48	48	4.00000	u

Source: Data obtained from the research work

DUNCAN COMPARISON TEST FOR ADDITIONS WITH FIBRE ISOLATED FROM CAÑIHUA BRAN.

ORDER OF MERIT	ADDITIONS (%)	AVERAGE	SIGNIFICANCE LEVEL
1	5	6.74479167	a
2	4	6.70833333	a
3	3	6.50520833	b

Source: Data obtained from the research work

AVERAGES AND STANDARD DEVIATIONS OF SENSORY ANALYSIS OF YOGURT ADDED WITH FIBER ISOLATED FROM KIWICHA BRAN.

ADDITION (%)	AVERAGE	STANDARD DEVIATION
5	6.74479167	± 1.4352
4	6.70833333	± 1.3957
3	6.50520833	± 1.1673

Source: Data obtained from the research work

ANOVA, USING THE RANDOMIZED COMPLETE BLOCK DESIGN (DBCA).

ANOVA OF THE SENSORY ANALYSIS OF YOGHURT ADDED WITH FIBRE ISOLATED FROM KIWICHA BRAN.

F. of V.	G.L.	S.C.	C.M.	Fc	Meaning
A: Judges	49	407.7838	8.32212	11.3577	**
B: Treatments	2	4.900833	2.45042	3.34423	
AxB Interaction	1	3.351237	3.35123	4.57358	
Experimental error	98	71.80750	0.73273		
Total	149	484.4920			

Source: Data obtained from the research work

DUNCAN COMPARISON TEST FOR THE ADDITIONS OF FIBRE ISOLATED FROM THE KIWICHA BRAN.

ORDER OF MERIT	ADDITIONS	AVERAGE	SIGNIFICANCE LEVEL
1	5	6.93229167	a
2	3	6.82291667	b
3	4	6.48958333	c

Source: Data obtained from the research work

DUNCAN COMPARISON TEST FOR JUDGES WHO TASTED YOGHURT WITH FIBRE ISOLATED FROM KIWICHA BRAN.

ORDER OF MERIT	JUDGES	AVERAGE	SIGNIFICANCE LEVEL
1	18	8.75000	a
2	4	8.41667	b
3	38	8.41667	b
4	28	8.41667	b
5	7	8.33333	c
6	14	8.08333	c
7	43	7.83333	d
8	12	7.75000	d
9	5	7.66667	e
10	1	7.66667	e
11	8	7.58333	ef
12	13	7.50000	f
13	6	7.50000	f
14	33	7.41667	g
15	24	7.25000	h
16	17	7.25000	h
17	29	7.16667	i
18	15	7.16667	i
19	47	7.08333	j
20	3	7.08333	j
21	32	7.00000	k
22	27	7.00000	k
23	26	7.00000	k
24	35	6.91667	kl
25	21	6.83333	l
26	45	6.50000	m
27	37	6.41667	n
28	22	6.41667	n
29	2	6.41667	n
30	39	6.16667	o
31	31	6.16667	o
32	16	6.16667	o
33	44	6.08333	p
34	41	6.08333	p
35	40	6.08333	p
36	25	6.00000	q
37	9	6.00000	q
38	23	6.00000	q
39	20	6.00000	q
40	34	5.83333	r
41	19	5.83333	r
42	11	5.83333	r
43	36	5.66667	s
44	10	5.41667	t
45	46	5.33333	u
46	30	5.25000	v
47	42	5.00000	w
48	48	4.16667	x

Source: Data obtained from the research work

DUNCAN COMPARISON TEST FOR THE INTERACTION OF ADDITIONS - FIBRE ISOLATED FROM QUINOA, KIWICHA AND CAÑIHUA BRAN.

ORDER OF MERIT	ADDITIONS (%)	ISOLATED FIBER	AVERAGE	SIGNIFICANCE LEVEL
1	Control	Control	7.04687500	a
2	5	Kiwicha	6.93229167	b
3	3	Kiwicha	6.82291667	c
4	5	Quinoa	6.79166667	c
5	5	Cañihua	6.74479167	d
6	4	Cañihua	6.70833333	d
7	3	Cañihua	6.50520833	e
8	4	Kiwicha	6.48958333	f
9	4	Quinoa	6.26562500	g
10	3	Quinoa	6.00520833	h

Source: Data obtained from the research work

ANNEX 5
COMPARISONS AND ANOVAS FOR THE PHYSICOCHEMICAL CHARACTERISTICS OF YOGHURT WITH FIBRE ISOLATED FROM QUINOA, KIWICHA AND CAÑIHUA BRAN. PUNO JANUARY-MARCH 2015

pH OF THE MOST ACCEPTABLE YOGHURT WITH ISOLATED FIBER FROM QUINOA SAVAGE. PUNO JANUARY - MARCH 2015

DAYS	N	Media	Standard deviation	Standard error	95% confidence interval for the mean		Minimum	Maximum
					Lower limit	Upper limit		
0	2	5,3100	,01414	,01000	5,1829	5,4371	5,30	5,32
7	2	5,2100	,01414	,01000	5,0829	5,3371	5,20	5,22
14	2	5,1700	,01414	,01000	5,0429	5,2971	5,16	5,18
21	2	5,1250	,00707	,00500	5,0615	5,1885	5,12	5,13
28	2	5,1100	,01414	,01000	4,9829	5,2371	5,10	5,12
35	2	5,1000	,00000	,00000	5,1000	5,1000	5,10	5,10
Total	12	5,1708	,07645	,02207	5,1223	5,2194	5,10	5,32

Source: Data obtained from the research work

ANOVA FOR THE pH OF THE MOST ACCEPTABLE YOGHURT WITH FIBER ISOLATED FROM QUINOA BRANCH. PUNO JANUARY - MARCH 2015

COMPARISON	Sum of squares	gl	Root mean square	F	Sig.
Between groups	,063	5	,013	89,565	,000
Within groups	,001	6	,000		
Total	,064	11			

Source: Data obtained from the research work

TESTS FOR THE pH OF THE MOST ACCEPTABLE YOGHURT WITH FIBER ISOLATED FROM QUINOA BRANCH. PUNO JANUARY - MARCH 2015

TESTS	DAYS	N	Subset for alpha = 0.05			
			1	2	3	4
Tukey Ba	35	2	5,1000			
	28	2	5,1100			
	21	2	5,1250			
	14	2		5,1700		
	7	2			5,2100	
	0	2				5,3100
Duncana	35	2	5,1000			
	28	2	5,1100			
	21	2	5,1250			
	14	2		5,1700		
	7	2			5,2100	
	0	2				5,3100
	Sig.		,089	1,000	1,000	1,000

Source: data obtained from the research work

ACIDITY OF THE MOST ACCEPTABLE YOGURT WITH FIBER ISOLATED FROM QUINOA BRAN. PUNO JANUARY - MARCH 2015

DAYS	N	Media	Standard deviation	Standard error	95% confidence interval for the mean		Minimum	Maximum
					Lower limit	Upper limit		
0	2	,7350	,00707	,00500	,6715	,7985	,73	,74
7	2	,7500	,00000	,00000	,7500	,7500	,75	,75
14	2	,7600	,00000	,00000	,7600	,7600	,76	,76
21	2	,7600	,00000	,00000	,7600	,7600	,76	,76
28	2	,7600	,00000	,00000	,7600	,7600	,76	,76
35	2	,7600	,00000	,00000	,7600	,7600	,76	,76
Total	12	,7542	,00996	,00288	,7478	,7605	,73	,76

Source: Data obtained from the research work

ANOVA FOR ACIDITY OF THE MOST ACCEPTABLE YOGURT WITH FIBER ISOLATED FROM QUINOA BRAN. PUNO JANUARY - MARCH 2015

Comparison	Sum of squares	gl	Root mean square	F	Sig.
Between groups	,001	5	,000	25,000	,001
Within groups	,000	6	,000		
Total	,001	11			

Source: Data obtained from the research work

TESTS FOR ACIDITY OF THE MOST ACCEPTABLE YOGURT WITH FIBER ISOLATED FROM QUINOA BRAN. PUNO JANUARY - MARCH 2015

TESTS	DAYS	N	Subset for alpha = 0.05		
			1	2	3
Tukey Ba	0	2	,7350		
	7	2		,7500	
	14	2		,7600	
	21	2		,7600	
	28	2		,7600	
	35	2		,7600	
Duncana	0	2	,7350		
	7	2		,7500	
	14	2			,7600
	21	2			,7600
	28	2			,7600
	35	2			,7600
	Sig.		1,000	1,000	1,000

Source: Data obtained from the research work

MOST ACCEPTABLE YOGURT PROTEINS WITH FIBER ISOLATED FROM QUINOA BRAN. PUNO JANUARY - MARCH 2015

DAYS	N	Media	Standard deviation	Standard error	95% confidence interval for the mean		Minimum	Maximum
					Lower limit	Upper limit		
0	2	4,0500	,02828	,02000	3,7959	4,3041	4,03	4,07
7	2	4,0900	,00000	,00000	4,0900	4,0900	4,09	4,09

14	2	4,1000	,00000	,00000	4,1000	4,1000	4,10	4,10
21	2	4,1200	,01414	,01000	3,9929	4,2471	4,11	4,13
28	2	4,1400	,00000	,00000	4,1400	4,1400	4,14	4,14
35	2	4,1400	,00000	,00000	4,1400	4,1400	4,14	4,14
Total	12	4,1067	,03420	,00987	4,0849	4,1284	4,03	4,14

Source: Data obtained from the research work

ANOVA FOR MOST ACCEPTABLE YOGURT PROTEINS WITH FIBER ISOLATED FROM QUINOA BRAN. PUNO JANUARY - MARCH 2015

COMPARISON	Sum of squares	gl	Root mean square	F	Sig.
Between groups	,012	5	,002	14,240	,003
Within groups	,001	6	,000		
Total	,013	11			

Source: Data obtained from the research work

TESTS FOR MORE ACCEPTABLE YOGHURT PROTEINS WITH ISOLATED FIBRE FROM QUINOA BRAN. PUNO JANUARY - MARCH 2015

TESTS			Subset for alpha = 0.05		
	DAYS	N	1	2	3
Tukey Ba	0	2	4,0500		
	7	2	4,0900	4,0900	
	14	2		4,1000	4,1000
	21	2		4,1200	4,1200
	28	2			4,1400
	35	2			4,1400
Duncana	0	2	4,0500		
	7	2		4,0900	
	14	2		4,1000	
	21	2		4,1200	4,1200
	28	2			4,1400
	35	2			4,1400
	Sig.		1,000	,066	,185

Source: Data obtained from the research work

MORE ACCEPTABLE YOGHURT FAT WITH FIBRE ISOLATED FROM QUINOA BRAN. PUNO JANUARY - MARCH 2015

			Standard		95% confidence interval for the mean			
DAYS	N	Media	deviation	Standard error	Lower limit	Upper limit	Minimum	Maximum
0	2	3,0050	,00707	,00500	2,9415	3,0685	3,00	3,01
7	2	3,0850	,00707	,00500	3,0215	3,1485	3,08	3,09
14	2	3,0950	,00707	,00500	3,0315	3,1585	3,09	3,10
21	2	3,1050	,00707	,00500	3,0415	3,1685	3,10	3,11
28	2	3,1200	,00000	,00000	3,1200	3,1200	3,12	3,12
35	2	3,1200	,00000	,00000	3,1200	3,1200	3,12	3,12
Total	12	3,0883	,04130	,01192	3,0621	3,1146	3,00	3,12

Source: data obtained from the research work

ANOVA FOR THE MOST ACCEPTABLE YOGURT FAT WITH FIBER ISOLATED FROM QUINOA BRAN. PUNO JANUARY - MARCH 2015

COMPARISON	Sum of squares	gl	Root mean square	F	Sig.
Between groups	,019	5	,004	111,400	,000
Within groups	,000	6	,000		
Total	,019	11			

Source: Data obtained from the research work

TESTS FOR THE MOST ACCEPTABLE YOGURT FAT WITH FIBER ISOLATED FROM QUINOA BRAN. PUNO JANUARY - MARCH 2015

TESTS			Subset for alpha = 0.05			
	DAYS	N	1	2	3	4
Tukey Ba	0	2	3,0050			
	7	2		3,0850		
	14	2		3,0950		
	21	2		3,1050	3,1050	
	28	2			3,1200	
	35	2			3,1200	
Duncana	0	2	3,0050			
	7	2		3,0850		
	14	2		3,0950	3,0950	
	21	2			3,1050	
	28	2				3,1200
	35	2				3,1200
	Sig.		1,000	,134	,134	1,000

Source: Data obtained from the research work

MORE ACCEPTABLE YOGURT WITH FIBER ISOLATED FROM QUINOA BRAN. PUNO JANUARY - MARCH 2015

					95% confidence interval for the mean			
DAYS	N	Media	Standard deviation	Standard error	Lower limit	Upper limit	Minimum	Maximum
0	2	,3000	,00000	,00000	,3000	,3000	,30	,30
7	2	,3000	,00000	,00000	,3000	,3000	,30	,30
14	2	,3000	,00000	,00000	,3000	,3000	,30	,30
21	2	,3000	,00000	,00000	,3000	,3000	,30	,30
28	2	,3000	,00000	,00000	,3000	,3000	,30	,30
35	2	,3350	,00707	,00500	,2715	,3985	,33	,34
Total	12	,3058	,01379	,00398	,2971	,3146	,30	,34

Source: Data obtained from the research work

ANOVA FOR THE MOST ACCEPTABLE YOGURT WITH FIBER ISOLATED FROM QUINOA BRAN. PUNO JANUARY - MARCH 2015

COMPARISON	Sum of squares	gl	Root mean square	F	Sig.
Between groups	,002	5	,000	49,000	,000
Within groups	,000	6	,000		
Total	,002	11			

Source: Data obtained from the research work

TESTS FOR THE MOST ACCEPTABLE YOGHURT WITH FIBRE ISOLATED FROM QUINOA BRAN. PUNO JANUARY - MARCH 2015

TESTS	DAYS	N	Subset for alpha = 0.05	
			1	2
Tukey Ba	0	2	,3000	
	7	2	,3000	
	14	2	,3000	
	21	2	,3000	
	28	2	,3000	
	35	2		,3350
Duncana	0	2	,3000	
	7	2	,3000	
	14	2	,3000	
	21	2	,3000	
	28	2	,3000	
	35	2		,3350
	Sig.		1,000	1,000

Source: Data obtained from the research work

pH OF THE MOST ACCEPTABLE YOGHURT WITH ISOLATED FIBER FROM CAÑIHUA SAVINGS. PUNO JANUARY - MARCH 2015

DAYS	N	Media	Standard deviation	Standard error	95% confidence interval for the mean		Minimum	Maximum
					Lower limit	Upper limit		
0	2	5,1600	,01414	,01000	5,0329	5,2871	5,15	5,17
7	2	5,1200	,00000	,00000	5,1200	5,1200	5,12	5,12
14	2	5,0850	,00707	,00500	5,0215	5,1485	5,08	5,09
21	2	5,0050	,00707	,00500	4,9415	5,0685	5,00	5,01
28	2	4,9000	,00000	,00000	4,9000	4,9000	4,90	4,90
35	2	4,9000	,00000	,00000	4,9000	4,9000	4,90	4,90
Total	12	5,0283	,10667	,03079	4,9606	5,0961	4,90	5,17

Source: Data obtained from the research work

ANOVA FOR THE pH OF THE MOST ACCEPTABLE YOGHURT WITH FIBER ISOLATED FROM CAÑIHUA SAVINGS, PUNO JANUARY - MARCH 2015

COMPARISON	Sum of squares	gl	Root mean square	F	Sig.
Between groups	,125	5	,025	499,467	,000
Within groups	,000	6	,000		
Total	,125	11			

Source: Data obtained from the research work

TESTS FOR THE pH OF THE MOST ACCEPTABLE YOGHURT WITH FIBER ISOLATED FROM CAÑIHUA BRANCH PUNO JANUARY - MARCH 2015

TESTS	DAYS	N	Subset for alpha = 0.05				
			1	2	3	4	5
Tukey Ba	28	2	4,9000				
	35	2	4,9000				
	21	2		5,0050			
	14	2			5,0850		
	7	2				5,1200	
	0	2					5,1600
Duncana	28	2	4,9000				
	35	2	4,9000				
	21	2		5,0050			
	14	2			5,0850		
	7	2				5,1200	
	0	2					5,1600
	Sig.		1,000	1,000	1,000	1,000	1,000

Source: Data obtained from the research work

ACIDITY OF THE MOST ACCEPTABLE YOGHURT WITH ISOLATED FIBRE FROM CAÑIHUA BRAN. PUNO JANUARY - MARCH 2015

DAYS	N	Media	Standard deviation	Standard error	95% confidence interval for the mean		Minimum	Maximum
					Lower limit	Upper limit		
0	2	,7050	,00707	,00500	,6415	,7685	,70	,71
7	2	,7160	,00141	,00100	,7033	,7287	,72	,72
14	2	,7200	,00000	,00000	,7200	,7200	,72	,72
21	2	,7350	,00707	,00500	,6715	,7985	,73	,74
28	2	,7500	,00000	,00000	,7500	,7500	,75	,75
35	2	,7600	,00000	,00000	,7600	,7600	,76	,76
Total	12	,7310	,02041	,00589	,7180	,7440	,70	,76

Source: Data obtained from the research work

ANOVA FOR THE ACIDITY OF THE MOST ACCEPTABLE YOGURT WITH FIBER ISOLATED FROM QUINOA BRAN. PUNO JANUARY - MARCH 2015

COMPARISON	Sum of squares	gl	Root mean square	F	Sig.
Between groups	,004	5	,001	52,706	,000
Within groups	,000	6	,000		
Total	,005	11			

Source: Data obtained from the research work

TESTS FOR THE ACIDITY OF THE MOST ACCEPTABLE YOGHURT WITH FIBRE ISOLATED FROM CAÑIHUA BRAN. PUNO JANUARY - MARCH 2015

TESTS	DAYS	N	Subset for alpha = 0.05			
			1	2	3	4
Tukey [Ba]	0	2	,7050			
	7	2	,7160	,7160		
	14	2		,7200		
	21	2			,7350	
	28	2				,7500
	35	2				,7600
Duncana	0	2	,7050			
	7	2		,7160		
	14	2		,7200		
	21	2			,7350	
	28	2				,7500
	35	2				,7600
	Sig.		1,000	,369	1,000	,051

Source: Data obtained from the research work

MOST ACCEPTABLE YOGHURT PROTEINS WITH ISOLATED FIBRE FROM CAÑIHUA BRAN. PUNO JANUARY - MARCH 2015

DAYS	N	Media	Standard deviation	Standard error	95% confidence interval for the mean		Minimum	Maximum
					Lower limit	Upper limit		
0	2	3,8250	,00707	,00500	3,7615	3,8885	3,82	3,83
7	2	3,8300	,00000	,00000	3,8300	3,8300	3,83	3,83
14	2	3,8400	,00000	,00000	3,8400	3,8400	3,84	3,84
21	2	3,8450	,00707	,00500	3,7815	3,9085	3,84	3,85
28	2	3,8500	,00000	,00000	3,8500	3,8500	3,85	3,85
35	2	3,8575	,00354	,00250	3,8257	3,8893	3,86	3,86
Total	12	3,8413	,01208	,00349	3,8336	3,8489	3,82	3,86

Source: Data obtained from the research work

ANOVA FOR THE MOST ACCEPTABLE YOGHURT PROTEINS WITH ISOLATED FIBRE FROM CAHIHUA BRAN. PUNO JANUARY - MARCH 2015

COMPARISON	Sum of squares	gl	Root mean square	F	Sig.
Between groups	,001	5	,000	15,933	,002
Within groups	,000	6	,000		
Total	,002	11			

Source: Data obtained from the research work

TESTS FOR THE MOST ACCEPTABLE YOGHURT PROTEINS WITH ISOLATED FIBRE FROM CAÑIHUA BRAN. PUNO JANUARY - MARCH 2015

TESTS			Subset for alpha = 0.05			
	DAYS	N	1	2	3	4
Tukey Ba	0	2	3,8250			
	7	2	3,8300	3,8300		
	14	2	3,8400	3,8400	3,8400	
	21	2		3,8450	3,8450	3,8450
	28	2			3,8500	3,8500
	35	2				3,8575
Duncana	0	2	3,8250			
	7	2	3,8300	3,8300		
	14	2		3,8400	3,8400	
	21	2			3,8450	
	28	2			3,8500	3,8500
	35	2				3,8575
	Sig.		,292	,060	,067	,134

Source: Data obtained from the research work

MORE ACCEPTABLE YOGHURT FAT WITH FIBRE ISOLATED FROM QUINOA BRAN. PUNO JANUARY - MARCH 2015

DAYS	N	Media	Standard deviation	Standard error	95% confidence interval for the mean		Minimum	Maximum
					Lower limit	Upper limit		
0	2	2,4950	,00707	,00500	2,4315	2,5585	2,49	2,50
7	2	2,5000	,00000	,00000	2,5000	2,5000	2,50	2,50
14	2	2,5950	,00707	,00500	2,5315	2,6585	2,59	2,60
21	2	2,6000	,00000	,00000	2,6000	2,6000	2,60	2,60
28	2	2,6000	,00000	,00000	2,6000	2,6000	2,60	2,60
35	2	2,6000	,00000	,00000	2,6000	2,6000	2,60	2,60
Total	12	2,5650	,05000	,01443	2,5332	2,5968	2,49	2,60

Source: Data obtained from the research work

ANOVA FOR THE MOST ACCEPTABLE YOGURT FAT WITH FIBER ISOLATED FROM CAÑIHUA BRAN. PUNO JANUARY - MARCH 2015

COMPARISON	Sum of squares	gl	Root mean square	F	Sig.
Between groups	,027	5	,005	328,800	,000
Within groups	,000	6	,000		
Total	,027	11			

Source: Data obtained from the research work

TESTS FOR THE MOST ACCEPTABLE YOGHURT FAT WITH FIBRE ISOLATED FROM CAÑIHUA BRAN. PUNO JANUARY - MARCH 2015

TESTS	DAYS	N	Subset for alpha = 0.05	
			1	2
Tukey [Ba]	0	2	2,4950	
	7	2	2,5000	
	14	2		2,5950
	21	2		2,6000
	28	2		2,6000
	35	2		2,6000
Duncana	0	2	2,4950	
	7	2	2,5000	
	14	2		2,5950
	21	2		2,6000
	28	2		2,6000
	35	2		2,6000
	Sig.		,267	,286

Source: Data obtained from the research work

MORE ACCEPTABLE YOGHURT WITH FIBRE ISOLATED FROM CAÑIHUA BRAN. PUNO JANUARY - MARCH 2015

DAYS	N	Media	Standard deviation	Standard error	95% confidence interval for the mean		Minimum	Maximum
					Lower limit	Upper limit		
0	2	,3000	,00000	,00000	,3000	,3000	,30	,30
7	2	,3000	,00000	,00000	,3000	,3000	,30	,30
14	2	,4000	,00000	,00000	,4000	,4000	,40	,40
21	2	,4000	,00000	,00000	,4000	,4000	,40	,40
28	2	,4000	,00000	,00000	,4000	,4000	,40	,40
35	2	,4550	,00707	,00500	,3915	,5185	,45	,46
Total	12	,3758	,05961	,01721	,3380	,4137	,30	,46

Source: Data obtained from the research work

ANOVA OF THE MOST ACCEPTABLE YOGHURT WITH ISOLATED FIBRE FROM CAÑIHUA BRAN. PUNO JANUARY - MARCH 2015

COMPARISON	Sum of squares	gl	Root mean square	F	Sig.
Between groups	,039	5	,008	937,000	,000
Within groups	,000	6	,000		
Total	,039	11			

Source: Data obtained from the research work

TESTING OF THE MOST ACCEPTABLE YOGHURT WITH ISOLATED FIBRE FROM CAÑIHUA BRAN. PUNO JANUARY - MARCH 2015

TESTS			Subset for alpha = 0.05		
	DAYS	N	1	2	3
Tukey [Ba]	0	2	,3000		
	7	2	,3000		
	14	2		,4000	
	21	2		,4000	
	28	2		,4000	
	35	2			,4550
Duncana	0	2	,3000		
	7	2	,3000		
	14	2		,4000	
	21	2		,4000	
	28	2		,4000	
	35	2			,4550
	Sig.		1,000	1,000	1,000

Source: Data obtained from the research work

pH OF THE MOST ACCEPTABLE YOGHURT WITH ISOLATED FIBER FROM KIWICHA SAVAGE. PUNO JANUARY - MARCH 2015

DAYS	N	Media	Standard deviation	Standard error	95% confidence interval for the mean		Minimum	Maximum
					Lower limit	Upper limit		
0	2	5,1100	,01414	,01000	4,9829	5,2371	5,10	5,12
7	2	5,1000	,00000	,00000	5,1000	5,1000	5,10	5,10
14	2	5,0000	,00000	,00000	5,0000	5,0000	5,00	5,00
21	2	4,8900	,01414	,01000	4,7629	5,0171	4,88	4,90
28	2	4,8000	,00000	,00000	4,8000	4,8000	4,80	4,80
35	2	4,7850	,00707	,00500	4,7215	4,8485	4,78	4,79
Total	12	4,9475	,13758	,03972	4,8601	5,0349	4,78	5,12

Source: Data obtained from the research work

ANOVA FOR THE pH OF THE MOST ACCEPTABLE YOGHURT WITH FIBER ISOLATED FROM KIWICHA SAVAGE. PUNO JANUARY - MARCH 2015

COMPARISON	Sum of squares	gl	Root mean square	F	Sig.
Between groups	,208	5	,042	554,067	,000
Within groups	,000	6	,000		
Total	,208	11			

Source: Data obtained from the research work

TESTS FOR THE pH OF THE MOST ACCEPTABLE YOGHURT WITH FIBER ISOLATED FROM KIWICHA SAVAGE. PUNO JANUARY - MARCH 2015

TESTS			Subset for alpha = 0.05			
	DAYS	N	1	2	3	4
Tukey Ba	35	2	4,7850			
	28	2	4,8000			
	21	2		4,8900		
	14	2			5,0000	
	7	2				5,1000
	0	2				5,1100
Duncana	35	2	4,7850			
	28	2	4,8000			
	21	2		4,8900		
	14	2			5,0000	
	7	2				5,1000
	0	2				5,1100
	Sig.		,134	1,000	1,000	,292

Source: Data obtained from the research work

MORE ACCEPTABLE YOGURT ACIDITY WITH FIBER ISOLATED FROM KIWICHA BRAN. PUNO JANUARY - MARCH 2015

			Standard		95% confidence interval for the mean			
DAYS	N	Media	deviation	Standard error	Lower limit	Upper limit	Minimum	Maximum
0	2	,7150	,02121	,01500	,5244	,9056	,70	,73
7	2	,7300	,00000	,00000	,7300	,7300	,73	,73
14	2	,7300	,00000	,00000	,7300	,7300	,73	,73
21	2	,7350	,00707	,00500	,6715	,7985	,73	,74
28	2	,7300	,00000	,00000	,7300	,7300	,73	,73
35	2	,7375	,00354	,00250	,7057	,7693	,74	,74
Total	12	,7296	,01010	,00292	,7232	,7360	,70	,74

Source: Data obtained from the research work

ANOVA FOR THE ACIDITY OF THE MOST ACCEPTABLE YOGURT WITH FIBER ISOLATED FROM QUINOA BRAN. PUNO JANUARY - MARCH 2015

COMPARISON	Sum of squares	gl	Root mean square	F	Sig.
Between groups	,001	5	,000	1,429	,335
Within groups	,001	6	,000		
Total	,001	11			

Source: Data obtained from the research work

MORE ACCEPTABLE YOGURT PROTEINS WITH FIBER ISOLATED FROM KIWICHA BRAN. PUNO JANUARY - MARCH 2015

DAYS	N	Media	Standard deviation	Standard error	95% confidence interval for the mean		Minimum	Maximum
					Lower limit	Upper limit		
0	2	3,2600	,01414	,01000	3,1329	3,3871	3,25	3,27
7	2	3,2800	,00000	,00000	3,2800	3,2800	3,28	3,28
14	2	3,2800	,00000	,00000	3,2800	3,2800	3,28	3,28
21	2	3,2950	,00707	,00500	3,2315	3,3585	3,29	3,30
28	2	3,3000	,00000	,00000	3,3000	3,3000	3,30	3,30
35	2	3,3750	,03536	,02500	3,0573	3,6927	3,35	3,40
Total	12	3,2983	,03996	,01154	3,2729	3,3237	3,25	3,40

Source: Data obtained from the research work

ANOVA FOR THE MOST ACCEPTABLE YOGHURT PROTEINS WITH FIBRE ISOLATED FROM KIWICHA BRAN. PUNO JANUARY - MARCH 2015

COMPARISON	Sum of squares	gl	Root mean square	F	Sig.
Between groups	,016	5	,003	12,853	,004
Within groups	,001	6	,000		
Total	,018	11			

Source: Data obtained from the research work

TESTS FOR THE MOST ACCEPTABLE YOGHURT PROTEINS WITH FIBRE ISOLATED FROM KIWICHA BRAN. PUNO JANUARY - MARCH 2015

TESTS	DAYS	N	Subset for alpha = 0.05	
			1	2
Tukey Ba	0	2	3,2600	
	7	2	3,2800	
	14	2	3,2800	
	21	2	3,2950	
	28	2	3,3000	
	35	2		3,3750
Duncana	0	2	3,2600	
	7	2	3,2800	
	14	2	3,2800	
	21	2	3,2950	
	28	2	3,3000	
	35	2		3,3750
	Sig.		,055	1,000

Source: Data obtained from the research work

MORE ACCEPTABLE YOGHURT FAT WITH FIBRE ISOLATED FROM KIWICHA BRAN. PUNO JANUARY - MARCH 2015

DAYS	N	Media	Standard deviation	Standard error	95% confidence interval for the mean		Minimum	Maximum
					Lower limit	Upper limit		
0	2	2,4000	,00000	,00000	2,4000	2,4000	2,40	2,40
7	2	2,4000	,00000	,00000	2,4000	2,4000	2,40	2,40
14	2	2,4000	,00000	,00000	2,4000	2,4000	2,40	2,40
21	2	2,4000	,00000	,00000	2,4000	2,4000	2,40	2,40
28	2	2,4000	,00000	,00000	2,4000	2,4000	2,40	2,40
35	2	2,5250	,03536	,02500	2,2073	2,8427	2,50	2,55
Total	12	2,4208	,04981	,01438	2,3892	2,4525	2,40	2,55

Source: Data obtained from the research work

ANOVA FOR THE MOST ACCEPTABLE YOGURT FAT WITH FIBER ISOLATED FROM KIWICHA BRAN. PUNO JANUARY - MARCH 2015

COMPARISON	Sum of squares	gl	Root mean square	F	Sig.
Between groups	,026	5	,005	25,000	,001
Within groups	,001	6	,000		
Total	,027	11			

Source: Data obtained from the research work

TESTS FOR THE MOST ACCEPTABLE YOGHURT FAT WITH FIBRE ISOLATED FROM KIWICHA BRAN. PUNO JANUARY - MARCH 2015

TESTS	DAYS	N	Subset for alpha = 0.05	
			1	2
Tukey Ba	0	2	2,4000	
	7	2	2,4000	
	14	2	2,4000	
	21	2	2,4000	
	28	2	2,4000	
	35	2		2,5250
Duncana	0	2	2,4000	
	7	2	2,4000	
	14	2	2,4000	
	21	2	2,4000	
	28	2	2,4000	
	35	2		2,5250
	Sig.		1,000	1,000

Source: Data obtained from the research work

MORE ACCEPTABLE YOGHURT WITH FIBRE ISOLATED FROM KIWICHA BRAN. PUNO JANUARY - MARCH 2015

DAYS	N	Media	Standard deviation	Standard error	95% confidence interval for the mean Lower limit	Upper limit	Minimum	Maximum
0	2	5,0600	,02828	,02000	4,8059	5,3141	5,04	5,08
7	2	5,0300	,01414	,01000	4,9029	5,1571	5,02	5,04
14	2	5,0000	,00000	,00000	5,0000	5,0000	5,00	5,00
21	2	4,8000	,00000	,00000	4,8000	4,8000	4,80	4,80
28	2	4,5000	,00000	,00000	4,5000	4,5000	4,50	4,50
35	2	4,5000	,00000	,00000	4,5000	4,5000	4,50	4,50
Total	12	4,8150	,24850	,07174	4,6571	4,9729	4,50	5,08

Source: Data obtained from the research work

ANOVA FOR THE MOST ACCEPTABLE YOGHURT WITH FIBRE ISOLATED FROM KIWICHA BRAN. PUNO JANUARY - MARCH 2015

COMPARISON	Sum of squares	gl	Root mean square	F	Sig.
Between groups	,026	5	,005	625,000	,000
Within groups	,000	6	,000		
Total	,026	11			

Source: Data obtained from the research work

TESTS FOR THE MOST ACCEPTABLE YOGHURT WITH FIBRE ISOLATED FROM KIWICHA BRAN. PUNO JANUARY - MARCH 2015

TESTS	DAYS	N	Subset for alpha = 0.05 1	2
Tukey Ba	14	2	,3000	
	21	2	,3000	
	28	2	,3000	
	35	2	,3050	
	0	2		,4000
	7	2		,4000
Duncana	14	2	,3000	
	21	2	,3000	
	28	2	,3000	
	35	2	,3050	
	0	2		,4000
	7	2		,4000
	Sig.		,150	1,000

Source: Data obtained from the research work

pH OF YOGHURT CONTROL. PUNO JANUARY - MARCH 2015

DAYS	N	Media	Standard deviation	Standard error	95% confidence interval for the mean		Minimum	Maximum
					Lower limit	Upper limit		
0	2	5,0600	,02828	,02000	4,8059	5,3141	5,04	5,08
7	2	5,0300	,01414	,01000	4,9029	5,1571	5,02	5,04
14	2	5,0000	,00000	,00000	5,0000	5,0000	5,00	5,00
21	2	4,8000	,00000	,00000	4,8000	4,8000	4,80	4,80
28	2	4,5000	,00000	,00000	4,5000	4,5000	4,50	4,50
35	2	4,5000	,00000	,00000	4,5000	4,5000	4,50	4,50
Total	12	4,8150	,24850	,07174	4,6571	4,9729	4,50	5,08

Source: Data obtained from the research work

ANOVA FOR YOGURT CONTROL. PUNO JANUARY - MARCH 2015

COMPARISON	Sum of squares	gl	Root mean square	F	Sig.
Between groups	,678	5	,136	813,960	,000
Within groups	,001	6	,000		
Total	,679	11			

Source: Data obtained from the research work

TESTS FOR YOGHURT pH CONTROL. PUNO JANUARY - MARCH 2015

TESTS	DAYS	N	Subset for alpha = 0.05			
			1	2	3	4
Tukeya	28	2	4,5000			
	35	2	4,5000			
	21	2		4,8000		
	14	2			5,0000	
	7	2			5,0300	5,0300
	0	2				5,0600
	Sig.		1,000	1,000	,310	,310
Duncana	28	2	4,5000			
	35	2	4,5000			
	21	2		4,8000		
	14	2			5,0000	
	7	2			5,0300	5,0300
	0	2				5,0600
	Sig.		1,000	1,000	,059	,059

Source: Data obtained from the research work

YOGURT ACIDITY CONTROL. PUNO JANUARY - MARCH 2015

DAYS	N	Media	Standard deviation	Standard error	95% confidence interval for the mean		Minimum	Maximum
					Lower limit	Upper limit		
0	2	,7600	,00000	,00000	,7600	,7600	,76	,76
7	2	,7600	,00000	,00000	,7600	,7600	,76	,76
14	2	,7700	,00000	,00000	,7700	,7700	,77	,77
21	2	,7750	,00707	,00500	,7115	,8385	,77	,78
28	2	,7750	,00707	,00500	,7115	,8385	,77	,78
35	2	,7800	,00000	,00000	,7800	,7800	,78	,78
Total	12	,7700	,00853	,00246	,7646	,7754	,76	,78

Source: Data obtained from the research work

ANOVA FOR YOGURT ACIDITY CONTROL. PUNO. JANUARY - MARCH 2015

COMPARISON	Sum of squares	gl	Root mean square	F	Sig.
Between groups	,001	5	,000	8,400	,011
Within groups	,000	6	,000		
Total	,001	11			

Source: Data obtained from the research work

TESTS FOR YOGURT ACIDITY CONTROL. PUNO JANUARY - MARCH 2015

TESTS	DAYS	N	Subset for alpha = 0.05	
			1	2
Tukeya	0	2	,7600	
	7	2	,7600	
	14	2	,7700	,7700
	21	2	,7750	,7750
	28	2	,7750	,7750
	35	2		,7800
	Sig.		,069	,271
Duncana	0	2	,7600	
	7	2	,7600	
	14	2	,7700	,7700
	21	2		,7750
	28	2		,7750
	35	2		,7800
	Sig.		,056	,059

Source: Data obtained from the research work

CONTROL YOGURT PROTEINS. PUNO JANUARY - MARCH 2015

DAYS	N	Media	Standard deviation	Standard error	95% confidence interval for the mean		Minimum	Maximum
					Lower limit	Upper limit		
0	2	3,3650	,00707	,00500	3,3015	3,4285	3,36	3,37
7	2	3,3650	,00707	,00500	3,3015	3,4285	3,36	3,37
14	2	3,3700	,00000	,00000	3,3700	3,3700	3,37	3,37
21	2	3,3950	,00707	,00500	3,3315	3,4585	3,39	3,40
28	2	3,4000	,00000	,00000	3,4000	3,4000	3,40	3,40
35	2	3,4000	,00000	,00000	3,4000	3,4000	3,40	3,40
Total	12	3,3825	,01712	,00494	3,3716	3,3934	3,36	3,40

Source: Data obtained from the research work

ANOVA FOR YOGURT PROTEIN CONTROL. PUNO JANUARY - MARCH 2015

COMPARISON	Sum of squares	gl	Root mean square	F	Sig.
Between groups	,003	5	,001	24,600	,001
Within groups	,000	6	,000		
Total	,003	11			

Source: Data obtained from the research work

TESTS FOR CONTROL YOGURT PROTEINS. PUNO JANUARY - MARCH 2015

TESTS	DAYS	N	Subset for alpha = 0.05	
			1	2
Tukeya	0	2	3,3650	
	7	2	3,3650	
	14	2	3,3700	
	21	2		3,3950
	28	2		3,4000
	35	2		3,4000
	Sig.		,902	,902
Duncana	0	2	3,3650	
	7	2	3,3650	
	14	2	3,3700	
	21	2		3,3950
	28	2		3,4000
	35	2		3,4000
	Sig.		,370	,370

Source: Data obtained from the research work

MESOPHILIC AEROBES IN THE MOST ACCEPTABLE YOGURT WITH FIBER ISOLATED FROM QUINOA BRAN. PUNO JANUARY - MARCH 2015

DAYS	N	Media	Standard deviation	Standard error	95% confidence interval for the mean		Minimum	Maximum
					Lower limit	Upper limit		
0	2	1625000,0000	35355,33906	25000,00000	1307344,8816	1942655,1184	1,60E+6	1,65E+6
35	2	1810000,0000	14142,13562	10000,00000	1682937,9526	1937062,0474	1,80E+6	1,82E+6
Total	4	1717500,0000	109048,91869	54524,45934	1543978,8358	1891021,1642	1,60E+6	1,82E+6

Source: Data obtained from the research work

ANOVA FOR MESOPHILIC AEROBES IN THE MOST ACCEPTABLE YOGURT WITH FIBER ISOLATED FROM QUINOA BRAN. PUNO JANUARY - MARCH 2015

COMPARISON	Sum of squares	gl	Root mean square	F	Sig.
Between groups	34225000000,000	1	34225000000,000	47,207	,021
Within groups	1450000000,000	2	725000000,000		
Total	35675000000,000	3			

Source: Data obtained from the research work

MESOPHILIC AEROBES IN THE MOST ACCEPTABLE YOGURT WITH FIBER ISOLATED FROM KIWICHA BRAN. PUNO JANUARY - MARCH 2015

DAYS	N	Media	Standard deviation	Standard error	95% confidence interval for the mean		Minimum	Maximum
					Lower limit	Upper limit		
0	2	455000,0000	49497,47468	35000,00000	10282,8342	899717,1658	420000,00	490000,00
35	2	765000,0000	21213,20344	15000,00000	574406,9290	955593,0710	750000,00	780000,00
Total	4	610000,0000	181659,02125	90829,51062	320939,9595	899060,0405	420000,00	780000,00

Source: Data obtained from the research work

ANOVA FOR MESOPHILIC AEROBES IN THE MOST ACCEPTABLE YOGURT WITH FIBER ISOLATED FROM KIWICHA BRAN. PUNO JANUARY - MARCH 2015

COMPARISON	Sum of squares	gl	Root mean square	F	Sig.
Between groups	96100000000,000	1	96100000000,000	66,276	,015
Within groups	2900000000,000	2	1450000000,000		
Total	99000000000,000	3			

Source: Data obtained from the research work

AEROBIC MESOPHILIC BACTERIA IN THE MOST ACCEPTABLE YOGHURT WITH FIBRE ISOLATED FROM CAÑIHUA BRAN. PUNO JANUARY - MARCH 2015

DAYS	N	Media	Standard deviation	Standard error	95% confidence interval for the mean		Minimum	Maximum
					Lower limit	Upper limit		
0	2	640000,0000	28284,27125	20000,00000	385875,9053	894124,0947	620000,00	660000,00
35	2	750000,0000	14142,13562	10000,00000	622937,9526	877062,0474	740000,00	760000,00
Total	4	695000,0000	66080,75867	33040,37934	589850,7669	800149,2331	620000,00	760000,00

Source: Data obtained from the research work

ANOVA FOR MESOPHILIC AEROBES IN THE MOST ACCEPTABLE YOGHURT WITH FIBRE ISOLATED FROM CAÑIHUA BRAN. PUNO JANUARY - MARCH 2015

COMPARISON	Sum of squares	Gl	Root mean square	F	Sig.
Between groups	12100000000,000	1	12100000000,000	24,200	,039
Within groups	1000000000,000	2	500000000,000		
Total	13100000000,000	3			

Source: Data obtained from the research work

MESOPHILIC AEROBES IN YOGURT WITHOUT FIBER (CONTROL). PUNO JANUARY - MARCH 2015

DAYS	N	Media	Standard deviation	Standard error	95% confidence interval for the mean		Minimum	Maximum
					Lower limit	Upper limit		
0	2	335000,0000	176776,69530	-125000,00000	1253275,5920	1923275,5920	210000,00	460000,00
35	2	805000,0000	21213,20344	15000,00000	614406,9290	995593,0710	790000,00	820000,00
Total	4	570000,0000	290172,36257	145086,18129	108271,0184	1031728,9816	210000,00	820000,00

Source: Data obtained from the research work

ANOVA FOR MESOPHILIC AEROBES IN YOGURT WITHOUT FIBER (CONTROL). PUNO JANUARY - MARCH 2015

COMPARISON	Sum of squares	gl	Root mean square	F	Sig.
Between groups	220900000000,000	1	220900000000,000	13,937	,065
Within groups	31700000000,000	2	15850000000,000		
Total	252600000000,000	3			

Source: data obtained from the research work

YEAST IN THE MOST ACCEPTABLE YOGURT WITH FIBER ISOLATED FROM QUINOA BRAN. PUNO JANUARY - MARCH 2015

DAYS	N	Media	Standard deviation	Standard error	95% confidence interval for the mean		Minimum	Maximum
					Lower limit	Upper limit		
0	2	36000,0000	2828,42712	2000,00000	10587,5905	61412,4095	34000,00	38000,00

35	2	530000,0000	2828,42712	2000,00000	504587,5905	555412,4095	528000,00	532000,00
Total	4	283000,0000	285220,38263	142610,19131	-170849,2764	736849,2764	34000,00	532000,00

Source: data obtained from the research work

ANOVA FOR YEASTS IN THE MOST ACCEPTABLE YOGURT WITH FIBER ISOLATED FROM QUINOA BRAN. PUNO JANUARY - MARCH 2015

COMPARISON	Sum of squares	Gl	Root mean square	F	Sig.
Between groups	244036000000,000	1	244036000000,000	30504,500	,000
Within groups	16000000,000	2	8000000,000		
Total	244052000000,000	3			

Source: data obtained from the research work

YEAST IN THE MOST ACCEPTABLE YOGURT WITH FIBER ISOLATED FROM KIWICHA BRAN. PUNO JANUARY - MARCH 2015

DAYS	N	Media	Standard deviation	Standard error	95% confidence interval for the mean		Minimum	Maximum
					Lower limit	Upper limit		
0	2	150000,0000	56568,54249	40000,00000	-358248,1894	658248,1894	110000,00	190000,00
35	2	425500,0000	2121,32034	1500,00000	406440,6929	444559,3071	424000,00	427000,00
Total	4	287750,0000	162383,03483	81191,51741	29362,3554	546137,6446	110000,00	427000,00

Source: data obtained from the research work

ANOVA FOR YEASTS IN THE MOST ACCEPTABLE YOGURT WITH FIBER ISOLATED FROM KIWICHA BRAN. PUNO JANUARY - MARCH 2015

COMPARISON	Sum of squares	Gl	Root mean square	F	Sig.
Between groups	75900250000,000	1	75900250000,000	47,371	,020
Within groups	3204500000,000	2	1602250000,000		
Total	79104750000,000	3			

Source: data obtained from the research work

YEASTS IN THE MOST ACCEPTABLE YOGURT WITH FIBER ISOLATED FROM CAÑIHUA BRAN. PUNO JANUARY - MARCH 2015

DAYS	N	Media	Standard deviation	Standard error	95% confidence interval for the mean		Minimum	Maximum
					Lower limit	Upper limit		
0	2	30500,0000	4949,74747	3500,00000	-13971,7166	74971,7166	27000,00	34000,00
35	2	427000,0000	1414,21356	1000,00000	414293,7953	439706,2047	426000,00	428000,00
Total	4	228750,0000	228938,67447	114469,33723	-135542,5193	593042,5193	27000,00	428000,00

Source: data obtained from the research work

ANOVA FOR YEASTS IN THE MOST ACCEPTABLE YOGHURT WITH FIBRE ISOLATED FROM CAÑIHUA BRAN. PUNO JANUARY - MARCH 2015

COMPARISON	Sum of squares	gl	Root mean square	F	Sig.
Between groups	157212250000,000	1	157212250000,000	11865,075	,000
Within groups	26500000,000	2	13250000,000		
Total	157238750000,000	3			

Source: data obtained from the research work

YEASTS IN THE MOST ACCEPTABLE YOGURT WITHOUT FIBER (CONTROL). PUNO JANUARY - MARCH 2015

DAYS	N	Media	Standard deviation	Standard error	95% confidence interval for the mean		Minimum	Maximum
					Lower limit	Upper limit		
0	2	49000,0000	1414,21356	1000,00000	36293,7953	61706,2047	48000,00	50000,00
35	2	805000,0000	21213,20344	15000,00000	614406,9290	995593,0710	790000,00	820000,00
Total	4	427000,0000	436649,36352	218324,68176	-267806,5768	1121806,5768	48000,00	820000,00

Source: data obtained from the research work

ANOVA FOR YEASTS IN YOGURT WITHOUT FIBER (CONTROL). PUNO JANUARY - MARCH 2015

COMPARISON	Sum of squares	Gl	Root mean square	F	Sig.
Between groups	571536000000,000	1	571536000000,000	2528,920	,000
Within groups	452000000,000	2	226000000,000		
Total	571988000000,000	3			

Source: data obtained from the research work

ANNEX 6
SHELF LIFE OF THE MOST ACCEPTABLE YOGHURT WITH FIBRE ISOLATED FROM QUINOA, KIWICHA AND CAÑIHUA BRAN

			5ºC	15ºC	25ºC (77ºF)
		DIAS	PH	PH	PH
		0	5.31	3.5	3.6
		7	5.21	3.5	3.7
		14	5.17	3.6	
		21	5.125	3.6	
Ci= Ai+Bt	A		5.293	3.49	3.6
	B=K		-0.0085	0.0057	0.0143
	R		0.9482	0.8	1
Ci= Ai.ek. T	A		5.2931	3.4902	3.6
	B=K		-0.002	0.0016	0.0039
	R		0.9504	0.8	1

The best equation is the exponential equation because it has a better R

Ci= $Ai.e^{-kt}$

The average value of Ai is 4.12776667

Calculation of the Kt value based on the Arrhennius equation

Log (k)= Log (ko)-$\frac{Ea}{2.3R}\cdot\frac{1}{Ta}$

T °C	K= (1/min)
5	0.002
15	0.0016
25	0.0039

1/T	Log k
0.00359712	-2.69897
0.00347222	-2.79588002
0.0033557	-2.40893539

The equation according to Arrhenius is: Log (kt) = 1.458- 1177.7 $\frac{1}{Ta}$

Calculating for a temperature of 5°C
Replacing in Arrhenius

Log(kt)= -2.77833094
kt= 0.00166598

The value of Kt found we replace it in Ci

Ci= $Ai.e^{-kt}$

Ci= 4.12776667.e-0.00166598t

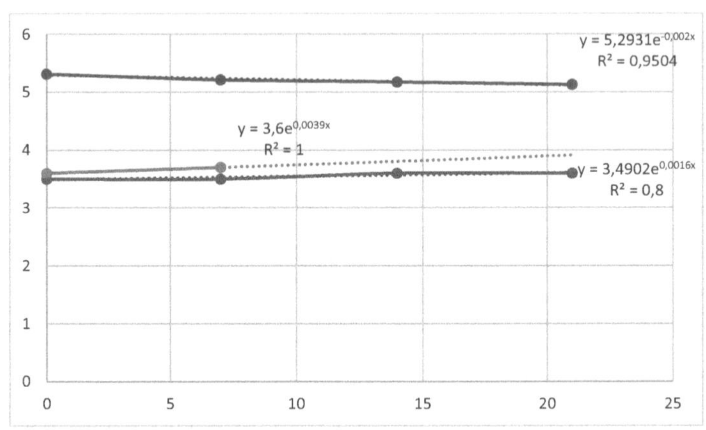

y = 5,2931e^-0,002x
R² = 0,9504
y = 3,6e^0,0039x
R² = 1
y = 3,4902e^0,0016x
R² = 0,8
6
5
4
3
2
1
0
0
5
10
15
20
25

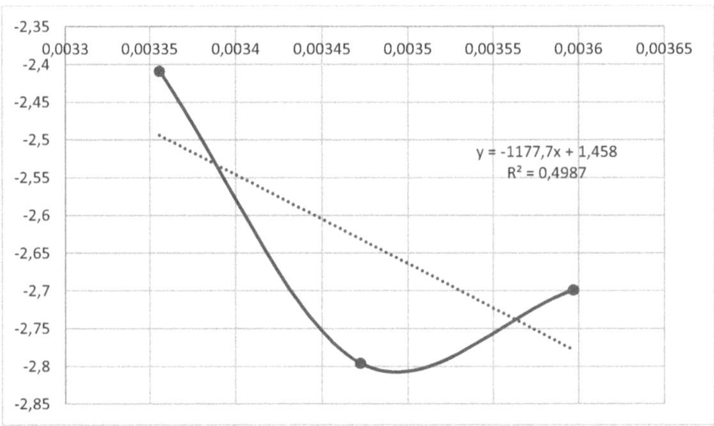

-2,35
-2,4
-2,45
-2,5
-2,55
-2,6
-2,65
-2,7
-2,75
-2,8
-2,85
0,0033
0,00335
0,0034
0,00345
0,0035
0,00355
0,0036
0,00365
y = -1177,7x + 1,458
R² = 0,4987

Printed by Books on Demand GmbH, Norderstedt / Germany